Uncertain Futures

Praise for this book

'This is a fascinating and timely book. It deals with a set of highly complex issues surrounding the practical dimensions of tackling community-based adaptation.'
Emily Boyd, School of Earth and Environment, University of Leeds

'By focusing on adaptive capacity and framing this in terms of resilience and complexity, Jonathan Ensor's new book sets out a powerful and challenging view of development in the age of climate change. With a possible future of four degrees warming, such radical approaches to development and adaptation are necessary, vital and urgent.'
Katrina Brown, Professor of Development Studies, University of East Anglia

'*Uncertain Futures* is a very welcome and valuable addition to the literature on climate change adaptation. The central contribution of this concise, thoughtful, and well-written book lies in its treatment of adaptive capacity as a key constitutive element of adaptation, but also as an imperative for a reorientation of development in which significantly more attention is given to the processes through which communities are empowered to create and shape change, rather than maintaining a focus on outcomes. In building his argument, Jonathan Ensor provides a fine explication of some very important concepts in the field of climate change adaptation that are often poorly understood or about which different interpretations exist. Central to this is adaptive capacity, but Ensor explores this in relation to other key concepts such as resilience and transformability, in the context of complex systems. *Uncertain Futures* marries intellectual rigour with lessons from development practice. In doing so, it presents a strong conceptual basis for designing and implementing high-quality development programmes focused on climate change adaptation. This book will appeal to development professionals, policymakers, and academic researchers, and is highly recommended.'
Alan Brouder, Head of Climate Change Adaptation, Oxfam GB

'This book will be an excellent guide for development policy makers, planners and practitioners grappling with incorporating climate change into development.'
Saleemul Huq, Senior Fellow, Climate Change Group at International Institute for Environment and Development

Uncertain Futures
Adapting development to a changing climate

Jonathan Ensor

Practical Action Publishing Ltd
25 Albert Street, Rugby, CV21 2SD, Warwickshire, UK
www.practicalactionpublishing.com

© Practical Action Publishing, 2011

ISBN 978 1 85339 720 2
ISBN Library Ebook: 9781780440392
Book DOI: http://dx.doi.org/10.3362/9781780440392

All rights reserved. No part of this publication may be reprinted or reproduced or utilized in any form or by any electronic, mechanical, or other means, now known or hereafter invented, including photocopying and recording, or in any information storage or retrieval system, without the written permission of the publishers.

A catalogue record for this book is available from the British Library.

The author has asserted his/her rights under the Copyright Designs and Patents Act 1988 to be identified as authors of this work.

Since 1974, Practical Action Publishing (formerly Intermediate Technology Publications and ITDG Publishing) has published and disseminated books and information in support of international development work throughout the world. Practical Action Publishing is a trading name of Practical Action Publishing Ltd (Company Reg. No. 1159018), the wholly owned publishing company of Practical Action. Practical Action Publishing trades only in support of its parent charity objectives and any profits are covenanted back to Practical Action (Charity Reg. No. 247257, Group VAT Registration No. 880 9924 76).

Cover photo: General view of the Azagarfa area, North Darfur, and desert terrain.
Credit: Mohammed Majzoub
Cover design by Practical Action Publishing
Indexed by Andrea Palmer
Typeset by S.J.I. Services, New Delhi

Contents

Preface	ix
About the author	xi
1. Community-based adaptation and development practice	**1**
Uncertain futures	2
Adaptation unpacked	3
Adaptive capacity and development practice	8
Adapting development	10
2. Understanding adaptive capacity	**13**
Resilience and adaptation	13
Defining adaptive capacity	15
Resilience thinking	18
Complexity	18
Thresholds and slow variables	19
Adaptive cycles	20
Resilience thinking for adaptive capacity	23
Dealing with complexity	25
Key messages from resilience thinking	28
3. Unpacking adaptive capacity	**31**
Supporting processes of change	31
Three dimensions of support for adaptive capacity	33
The three dimensions – unpacked	36
Power sharing	36
Knowledge and information	43
Experimentation and testing	46
Supporting adaptive capacity	48
4. Adapting development – in practice	**51**
Rights-based approaches: lessons for adaptive capacity	51
Lesson 1: participation should be transforming	52
Lesson 2: adaptive capacity is a political issue	54
Lesson 3: the state must be accountable	55
Lesson 4: the law may be a tool for adaptive capacity	56
Consensus building: sharing power and knowledge	58
Participatory Action Plan Development (PAPD)	58

Case study: PAPD in the Bangladesh charlands	63
Pathways to power sharing	68
Participatory technology development: experimentation and testing in practice	74
Case study: participatory technology development in Zimbabwe	76
Power-sharing approaches and PTD	79
Adapting development practice	81
5. Uncertain futures	85
Livelihood transitions and transformations	87
Living in the future	89
Adapting development	95
References	97
Index	105

Boxes

1.1	Climate modelling and forecasting: understanding the limits	4
3.1	Adaptive capacity and human development	32
3.2	Gender and adaptation decision making	37
3.3	The role of forecasting information	45
4.1	The six steps of PAPD	60
4.2	Possible stakeholders in natural resource management	63
4.3	The Mvoti stakeholder dialogue	69
4.4	DRR planning in different contexts: Nepal and Zimbabwe	71
4.5	Prajateerpu: a citizens' jury/scenario workshop on food and farming futures in Andhra Pradesh, India	73
4.6	Building confidence for experimentation	75
4.7	Sharing knowledge: seed fairs in Chivi	78
5.1	Cattle raiding in East Africa	90
5.2	Informal institutions for conflict resolution and local decision-making	92
5.3	Changing migration patterns	93

Figures

2.1	An adaptation perspective on changes to a social-ecological system following a disturbance	16
2.2	The adaptive cycle	21
2.3	The adaptive cycle and options for adaptation	24
3.1	A framework identifying areas for action on adaptive capacity	35
4.1	The U-process metaphor applied to the Mvoti dialogue	70
4.2	Towards self-sustaining support for adaptive capacity	82

Tables

1.1	Adaptation broken down into three components, each of which may be present in a combination determined by the local context	6
1.2	Examples of the three components of adaptation	6
2.1	Summary of resilience definitions	14
2.2	Shift in thinking and perspective for complex systems	19
3.1	The implications of resilience thinking for adaptive capacity	34

3.2	Steps towards power sharing	41
4.1	Pattern analysis for a hypothetical fishing community	61
4.2	Issues raised by different interest groups	65
4.3	Management requirements for the community action plan	66
4.4	Potential synergies between PAPD and PTD	79

Preface

From the late 1990s, rural communities were informing many international development agencies that they were experiencing much greater variability in the weather and the timing of seasons. Extreme weather events were more frequent and more severe, so that coping strategies developed over many generations were no longer adequate. Development practitioners have been adjusting to this new reality ever since. Practical Action, for example, initiated its first programme of work explicitly focused on community-based adaptation in 2004; adopted climate change as one of two cross-organizational goals in 2007; and in 2009 summarized five years of learning in a new book – *Understanding Climate Change Adaptation: lessons from community-based approaches*.

Since the publication of *Understanding Climate Change Adaptation* development organizations have continued to explore how programmes can reflect the need for communities to be able to adapt. Internationally, policy work has influenced the adaptation text in the UN Framework Convention on Climate Change (UNFCCC) negotiations and brought the realities of community-based adaptation to the UN's advisory mechanisms. Away from the international arena, development practitioners around the world have sought to bring the lessons of project work to the attention of developing country policy makers, while meetings in developed countries have stimulated discussion and shared learning between the different agencies working on community-based adaptation.

Adaptive capacity, presented in *Understanding Climate Change Adaptation* as a central component of adaptation, has emerged as a recurrent theme in these exchanges. For many, it represents what is new and different about development discourse in an age of climate change. But initial enthusiasm has given way to an important debate over what constitutes adaptive capacity and what it means for development practice. Practical Action has been actively engaged in this discussion, running several workshops in 2010 and early 2011 to explore the issues in greater depth. This book is intended to be a further contribution to this debate, in particular by proposing concrete actions that development policy and practice can – and, it is argued, must – support.

While this book has only been possible because of conversations with many colleagues over several years, some individuals have been invaluable in shaping its content. Julian Yates, now at the University of British Columbia, provided many thought-provoking responses to early drafts. His primary research in Nepal, undertaken on behalf of Practical Action, has been particularly valuable. Rachel Berger, Maggie Ibrahim and Barnaby Peacocke at Practical Action, and an anonymous reviewer, have all had a significant role in the development of the text, while Toby Milner, at Practical Action Publishing, was sufficiently enthusiastic about the ideas to ensure that writing got underway to begin with. Of course, the views and any errors within are the author's alone.

About the author

Jonathan Ensor works for Practical Action as a researcher, with particular interests in the relationship between climate change and international development. Jonathan has a background in development and human rights organizations, and his previous publications include *Understanding Climate Change Adaptation – Lessons from community-based approaches* (with Rachel Berger, Practical Action Publishing) and *Reinventing Development? Translating rights-based approaches from theory into practice* (co-edited with Paul Gready, Zed Books).

CHAPTER 1
Community-based adaptation and development practice

Community-based adaptation has emerged as an important new topic in international development. Its distinctive approach, in which responses to climate change are locally developed by the affected population, is increasingly understood as necessary to support those whose livelihoods and lives are most threatened by climate change. Today, intergovernmental bodies and non-governmental organizations alike are engaged in community focused adaptation projects, funded by national and international donors who recognize the significance and urgency of the challenge being faced. These initiatives are supported by a growing body of networks, conferences and publications. Among these, the Community Based Adaptation Exchange online resource centre lists more than 700 members from development, environment and research organizations, while the 2010 International Conference on Community-based Adaptation drew more than 170 delegates from 38 countries, and gave rise to a new global initiative to 'promote the exchange of knowledge about community-level adaptation to climate change'. A year later, at the 2011 conference numbers swelled to more than 300. Community-based adaptation, a topic that was more or less unknown at the turn of the century, now occupies an important place in both climate change and development debates.

Yet for all the rapid growth in interest, community-based adaptation remains a new concept whose meaning is still to be fully understood. Most interpretations agree on the need to increase community understanding of the challenges that climate change represents and to develop responses that build on local knowledge. Awareness raising and participatory approaches are combined with exposure to new livelihood or risk reduction strategies, enabling communities to develop locally appropriate solutions to locally defined problems. The specifics of implementation necessarily vary according to the context and draw on different pools of knowledge and experience. Communities faced with an increased risk of flooding, for example, require access to very different information and technologies to those faced with the emergence of new pests and diseases, or the gradual loss of dependable water sources. This process of supporting communities to develop responses to the challenges they face is a crucial step, and in some interpretations is the goal of community-based adaptation. In these cases, supporting the mechanisms through which communities create changes to their lives and livelihoods is very much a secondary concern to the provision of appropriate adaptation outcomes.

The premise of this book is different. The starting point is that community-based adaptation must be defined by a balance between actions that support the ongoing ability to change, and those that respond to current challenges. While meeting immediate needs will always be a priority, it should not become an end in itself. Rather, current challenges offer an entry point for actions that build the capacity to adapt: the overarching objective is to support the ability of communities to face an uncertain future. Reflecting a growing consensus, *adaptive capacity* in this book describes the component of an adaptation intervention that prepares communities for a changing climate, enabling them to better secure their own adaptation outcomes.

In community-based adaptation, adaptive capacity draws attention to the processes through which communities are able to make changes to their lives and livelihoods in response to emerging environmental change. As such, adaptive capacity challenges development actors to think in terms of how networks of relationships define the distribution, access and control of material and knowledge assets. It means that the quality of relationships, determined by characteristics such as power, culture and gender, are drawn into the foreground so that interventions can identify the constraints on local decision making, looking across scales rather than at communities in isolation. This perspective demands a shift in thinking away from needs-based programming, having more in common with empowerment-focused development narratives such as rights-based approaches. The increasing risk of four degrees of global warming suggests that frequent livelihood changes and even radical transformations may become necessary for many. In these circumstances, a business as usual approach to development practice is inadequate, and the necessity of securing adaptive capacity is thrown into sharp relief. The alternative to empowering communities to engage in processes of change is to leave them permanently dependent on the whim of governments who have routinely failed the most vulnerable, or on non-governmental actors whose funding and mandate cannot be relied on to provide indefinite support. The purpose of this book is to respond to this challenge, by addressing the need to better understand adaptive capacity, and to rethink development practice in terms of how it can be best supported.

Uncertain futures

At the heart of adaptation lies a paradox. The driving force behind much of the interest in adaptation is the knowledge that climate change is upon us and that the impacts will be felt, first and worst, in the world's most vulnerable communities. Yet, at the same time, these anticipated impacts are poorly understood, and the uncertainty is greatest in many of the world's least developed countries.

While advances in climate science mean that we now have an unprecedented view of the future of the earth system, many of the precise implications remain unclear: predictions of rainfall rates, the likely frequency of extreme weather events, and regional changes in weather patterns cannot be made with certainty (Ensor and Berger, 2009a). Box 1.1 summarizes the multiple forms of uncertainty that make predicting the impacts of climate change so difficult. But the presence of uncertainty should not be confused with a lack of knowledge: the issue is to develop a clear understanding of what climate science is offering. The impact of greenhouse gas emissions is well established and warming of the global climate throughout the coming century is certain. Climate change is already being experienced in many parts of the world, not least through rising sea levels, and climate models are clear in anticipating change at an unprecedented pace and scale. As time passes the emerging science continues to suggest that the changes may be more profound and with us sooner than first thought (Anderson and Bows, 2011).

However, while the global average temperature change can be anticipated with reasonable confidence for a given rate of future greenhouse gas emissions, predicting precipitation patterns and localized weather impacts at decadal to centennial timescales is beyond reach. In other words: 'The take-home message for policy makers is that for regions as small as most countries, knowing the global mean temperature leaves significant uncertainty in the local response. ... [O]ur knowledge of vulnerability to natural variability comes more from observations and science than from modelling.' (Smith, 2009: 145)

The established approach to accounting for uncertainty in climate models is to generate 'ensemble' predictions, in which the range of outputs provided by several different climate models are compared. Yet for some scientists, even the diversity of results obtained from all currently available models fails to fully describe future uncertainty, even for fixed future emissions (Stainforth et al., 2007; Smith, 2009). This is of central importance to adaptation. Whilst mitigation activities are rightly driven by the need to avoid dangerous climate change, adaptation planning relies on understanding what climate change means in a particular location. As a result, it is all too easy to assume that adaptation can and should follow climate change predictions. But climate models are not 'truth machines' that provide predictions (Desai et al., 2009: 70). Rather, they are increasingly refined simulations that are ultimately limited by our understanding of climate (and computing) science. In many contexts there is no agreement whether, for example, rainfall is likely to increase or reduce. What, then, should adaptation to climate change mean in these circumstances? How can adaptation planning proceed in the face of such uncertainty?

Adaptation unpacked

First and foremost, adaptation must avoid becoming an exercise in designing optimal solutions to anticipated climate change impacts. Options for changes to lives and livelihoods need to be robust in the face of unpredictable future weather patterns by ensuring that altered livelihood strategies do not only bring benefit if climate change plays out as predicted (Boyd et al., 2009).

4 UNCERTAIN FUTURES

Box 1.1 Climate modelling and forecasting: understanding the limits

Knowledge generation and the IPCC

The Intergovernmental Panel on Climate Change (IPCC) provides the most well-known and authoritative assessments of the current scientific understanding of climate change. Each report examines data from previously published peer reviewed literature (and selected non-peer reviewed reports) and is itself subjected to two rounds of expert review and one of government challenge and approval prior to publication. This approach removes controversial or spurious data and establishes a high degree of confidence in the content of the IPCC's publications. However, when relying on the IPCC's conclusions it is important to note that the approach to knowledge gathering and sharing has the potential to be conservative. Consensus building and time constraints mean that some evidence is excluded. For example, the IPCC's fourth report only considers temperature projections that fall within a 90 per cent confidence interval, excludes dynamic melting of the Greenland and Antarctic ice sheets, and excludes non-linear events that might result in higher or more rapid temperature or sea-level rise. These decisions are intended to prevent the IPCC from reporting speculative data and allow scientifically plausible statements of certainty to be made in one report – but they also mean that low probability, high-impact events are not drawn out for public consideration. Moreover, phenomena that cannot be modelled are ignored (Oppenheimer et al., 2007). The periodic release of the IPCC assessment reports is also an important limitation: the time taken by the IPCC review process means that the evidence relied on in the reports is restricted to that published well in advance of the IPCC release date, whilst the five to six year periodicity of reports mean that the most recent IPCC assessment lags behind current scientific thinking. Rather than undermining the IPCC conclusions or the importance of the process, these observations highlight the need to understand the limits that inevitably exist to even the most authoritative statements of knowledge.

Future emissions

Fundamentally, the relationship between human activity and climate change means that assumptions must be made about the pattern of future emissions in order to generate climate predictions. Climate change predictions are mainly dependent on the greenhouse gas composition of the atmosphere in the future (predominantly carbon dioxide, methane and nitrous oxide). This problem, also referred to as 'forcing uncertainty', is addressed by making multiple projections relating to a range of reasonably foreseeable future emissions.

Forecasting and modelling

The mechanisms involved in producing forecasts introduce uncertainty. Climate projections are significantly different from their more established cousin, weather forecasting. Climate refers to long-term (conventionally 20 to 30 year) average weather conditions. Weather, on the other hand, refers to short-term (hourly and daily) changes such as in temperature, rainfall and wind. Weather is hard to predict as its dynamics are chaotic: small changes in the current weather conditions can create large changes in the weather at a later time. Despite this, well-established scientific understanding and measurement infrastructure allows predictions up to about 15 days ahead.

Seasonal forecasting lies between weather and climate timescales and is based on slowly changing phenomena that have a significant impact on the weather, such as the El Niño Southern Oscillation (ENSO). Measuring these important but slowly changing phenomena allows seasonal trends to be predicted up to around two years in advance, although confidence is greater for shorter timescales of up to around three months. A typical seasonal forecast may predict daily rainfall for a particular three-month period. An expression of confidence is normally included: for example, a forecast may predict with

90 per cent confidence that daily rainfall will be between 150 mm and 200 mm. The confidence captures and communicates the uncertainty in the forecast, and varies significantly with geographical location: the more weather is dominated by El Niño, for example, the more accurate the seasonal forecast. Generally speaking, predictability reduces the further a location is from the equator and from the ocean, and temperature is usually easier to predict than precipitation.

Beyond seasonal timescales, climate models are relied on to provide information on long-term trends. Whilst they are able to establish with high confidence that global average temperatures will continue to increase (not least due to the levels of greenhouse gases currently in the atmosphere), more detailed changes, such as the impact of warming on wet and dry seasons, remain unclear. The IPCC's calibrated expressions of confidence draw attention to the inherent uncertainty in climate models, which by definition are only approximations of reality, offering an incomplete representation of the full complexity of the earth system. Even for a fixed rate of future emissions there is uncertainty as to the exact impact on temperature. Alongside model uncertainty, the initial conditions that are assumed or measured influence the outcome of projections (such as distribution of sea salinity, which can influence decadal-scale climate variations). Currently, the impact of uncertainty can be seen most clearly in the failure of climate models to provide good agreement at the regional (sub-continental) scale, and in particular on future levels of precipitation.

Impacts

Moving from climate model outputs to a prediction of the likely impacts of climate change requires a further layer of modelling, and therefore a further layer of uncertainty. Impact models are a separate area of study from climate science that rely on physical and socio-economic models to translate a climate future (changes in temperature, rainfall and length of growing season, for example) into human impacts (such as health implications, flooding and food supply). Such an assessment of impacts inevitably compounds uncertainties.

Tailoring smallholder agricultural strategies to a particular climate future, for example, risks maladaptation and will always be a mistake as long as uncertainty remains in climate projections. Rather, adaptation needs to tackle uncertainty head-on by ensuring that livelihoods retain and enhance the ability to ride out or respond to unexpected events.

Efforts to reduce current vulnerability to climate change impacts may form part of an adaptation strategy, but should be employed in combination with measures that specifically address future uncertainty. This can be achieved through a three-pronged approach, introduced in an analysis of eight early examples of community-based adaptation published in *Understanding Climate Change Adaptation* (Ensor and Berger, 2009a), and summarized here in Table 1.1 and explored in Table 1.2.

In this view, vulnerability reduction is an important component of adaptation to climate change for several reasons. In some cases, such as sea-level rise or glacial lake outburst flood risks, existing climate change threats can be clearly identified and adaptation measures that reduce vulnerability to the threat are a priority. But in most circumstances, establishing an unambiguous causal link to climate change may be impossible as attribution of weather related phenomenon is hugely problematic. By definition, climate is the average of weather conditions over a multi-decadal timescale; individual events or even multi-year phenomenon such as persistent drought can at best be described as consistent with anticipated climate change. Actions that reduce vulnerability to these threats are part of adaptation in the sense that they address existing challenges that are anticipated to persist or recur into the future. Often taking the form of technical solutions, they deal with conditions that are assumed to be linked to climate change.

Table 1.1 Adaptation broken down into three components, each of which may be present in a combination determined by the local context

Approach	Observations
Vulnerability reduction	• Vulnerability to climate change is assessed in reference to a particular hazard, for example vulnerability to flooding, and considers underlying human and environmental factors • Vulnerability reduction targets a particular hazard, and should aim to be 'no regrets': meeting short-term needs independent of future climate change
Strengthening absorbing capacity	• Defined by the ability to absorb shocks or ride out changes • Reduces vulnerability to a wide range of hazards • Supported by diversity of assets or livelihood strategies
Building adaptive capacity	• Defined by the ability to shape, create or respond to change • Enables measures to strengthen absorbing capacity and reduce vulnerability • Amount, diversity and distribution of assets facilitates alternative strategies • Requires information plus the capacity and opportunity to learn, experiment and make decisions

Source: adapted from Ensor and Berger, 2009a

Table 1.2 Examples of the three components of adaptation

Vulnerability reduction capacity *Targeting identified climate change vulnerability*	Strengthening absorbing capacity *Increasing the ability to absorb shocks or ride out changes*	Building adaptive capacity *Improving the ability to shape, create or respond to changes*
• Improving food security through new agricultural strategies • Securing domestic and agricultural water access via sand dams • Reducing coastal erosion and land loss through coastal planting	• Planting a diversity of seed types to address uncertainty in weather forecasts • Seed bulking and saving to secure access to traditional productive resources • Introduction of alternative livelihood practices	• Relationship building between community groups, local government and meteorological office • Training and support for experimenting with crop variety selection • Enabling participatory research and technology development

Source: case studies profiled in Ensor and Berger, 2009a

The natural variability in weather conditions and the inevitable uncertainty over the direction of future climate change mean that attempts at vulnerability reduction should focus on 'no regrets' strategies whose short-term benefit is maintained regardless of future prevailing weather conditions. Irrespective of attribution, such interventions also provide an entry point for those seeking to work with communities on the less tangible aspects of climate change. When addressed through community-based actions that raise awareness of the future challenges of climate change, existing vulnerabilities offer the opportunity to introduce an appreciation of climate risk in how current challenges are met. This approach embodies the maxim that community-based adaptation is not so much about what a community is doing, but why and with what knowledge (Huq, 2007). And when documented and shared, such actions also expand the global pool of knowledge on the climate-related challenges that poor communities face. This collection and redistribution of knowledge is increasingly recognized as an important precondition of effective adaptation as climate change brings new and unexpected challenges the world over.

While vulnerability reduction may be able to provide relief from climate-related challenges that are being experienced in the immediate term, a key question remains: how should adaptation deal with the problem of an uncertain future climate? What does a response to climate change that addresses as yet unknown impacts look like? The answer lies in shifting the focus of adaptation towards dealing with uncertainty directly (Boyd et al., 2009; Boyd and Juhola, 2009). This means working with communities in ways that strengthen their ability to cope with and recover from surprises, and strengthen their capacity to make changes to their lives as our knowledge of climate change and its likely impacts improves. Here, these strategies are referred to as strengthening absorbing capacity (the ability to absorb or cope with unexpected disturbances, often termed 'resilience') and adaptive capacity (the ability to change in response to climate change). These terms, adapted from Ensor and Berger (2009a), are explored in more depth in chapter 2.

Adaptive and absorbing capacity are not independent of vulnerability: as the examples in Table 1.2 suggest, increasing a household or community's ability to cope or adapt should help reduce vulnerability to the broadest possible range of hazards. Similarly, 'no regrets' approaches to vulnerability reduction frequently also increase the ability to absorb disturbances. For example, the introduction of multiple livelihood strategies or a shift in planting practices to include a diversity of varieties are both strategies that address the vulnerability of rural communities to the failure of crops due to reduced rainfall rates. Yet by increasing the diversity of income generating activities they also address a wide variety of future conditions. This is a result that emerges because 'no regrets' vulnerability reduction includes an appreciation of uncertainty, without which vulnerable livelihood practices may just as easily be replaced with a strategy that is best suited to the current climate, such as where saline intolerant crops, for example, are replaced with varieties whose yields are only able to cope with the current levels of salinity.

Adaptive capacity and development practice

The premise of this book is that adaptive capacity, while now part of the development lexicon, has still to be fully explored in terms of its meaning for development practice. Here, as elsewhere, it captures the ability to carry out adaptation measures. These may be aimed at reducing vulnerability to specific weather-related threats (a new seawall, adoption of heat and drought-tolerant crop varieties, vaccines, upgraded drainage systems) or at building livelihoods better able to cope with and recover from unexpected events (crop insurance schemes, agricultural diversification, enhanced water use efficiency). The vulnerability of communities to future climate change therefore depends in part on their adaptive capacity, through which timely adaptation measures can be adopted. This perspective encompasses not only the knowledge and resources that a community is able to deploy in adapting to climate change, but also the broader institutional and policy environment that supports or constrains decision making at the local level. While community-based adaptation rightly draws attention to the needs of the most vulnerable communities, the lens of adaptive capacity also locates those communities within a broader context, suggesting that action on adaptation will not only be necessary at the local level.

This understanding of the components of adaptation presents a particular challenge. Vulnerability and absorbing capacity are familiar territory for those concerned with development and disaster risk reduction. Working with communities to identify existing vulnerabilities and enhance the ability to cope with and recover from shocks is at the heart of many development organizations' praxis. This provides secure foundations of knowledge and experience from which to address the emerging challenges of climate change through community-based adaptation. But it also carries a risk that adaptation projects will be predicated on familiar methods, taking the need to reduce existing vulnerability to climate risks as the major challenge to be addressed. Strategies for adaptive capacity may then follow from this focus, but this is not the same as identifying and addressing in their own right the local barriers to adaptation.

In a case study representative of the majority considered in *Understanding Climate Change Adaptation* (Ensor and Berger, 2009a), a Sri Lankan project focused on the failure of hybrid rice varieties following saline intrusion. The approach taken aimed to revive consistently poor yields (vulnerability reduction). The method employed variety diversity (thereby strengthening the ability to cope with environmental disturbances) whilst developing the skills and opportunities needed to productively employ new varieties (building adaptive capacity). On one level, this approach satisfies the need to address uncertainty by including absorbing and adaptive capacity building measures. But by starting from known vulnerabilities, the intervention skewed the purpose of adaptation, with work focused on the well-understood challenge of variety access and selection, driven by the need for vulnerability reduction. The realities of future change were overlooked and thus adaptive capacity building measures emerged as a consequence of vulnerability reduction – building skills in employing new varieties – rather than through an independent focus on the measures necessary for meeting uncertainty. Climate change impacts other than salinity were not considered and therefore access to forecasting information (and an appreciation of its uncertain and statistical nature) were not perceived to be an essential step towards enabling future adaptation actions. This fundamental drawback was evident in several of the case studies and led to new agricultural practices and infrastructure being undermined by unanticipated climate variability, betraying a 'business as usual' mindset in which current conditions are assumed to be a sufficient indication of the future.

While the emphasis that is placed on each of the components of adaptation will vary according to local needs, a lack of clarity around adaptive capacity means that it can easily be overlooked or underplayed. In reviewing the challenges that climate change presents to the development community, Mitchell and Tanner (2006: 33) reflect that 'there is limited experience to date in combining measures that manage and reduce present-day risks but are suitably flexible and robust to cope with an uncertain future climate.' Nelson et al.'s review paper provides a possible explanation: 'analytically adaptation is most often narrowly conceptualized as a set of technological or technical options to respond to specific risks' (2007: 396) – in other words, vulnerability reduction. While academic literature has for some time discussed adaptive capacity (Folke et al., 2003; Smit and Wandel, 2006; Plumber and Armitage, 2010), it is only recently that publications from major development organizations have identified adaptive capacity as a component of adaptation. As a result, there remains a shortage of development experiences that explore its meaning, and definitions vary considerably. For example, CARE (2010) refer to the ability 'to moderate potential damages, take advantage of opportunities or cope with the consequences of climate change', Oxfam (Pettengell, 2010) talk in terms of the ability to be actively involved in processes of change, and the Swedish Commission on Climate Change and Development (2009), at the highest level of generality, correlate adaptive capacity with human development.

Each of these approaches to defining adaptive capacity is useful as they offer different perspectives on meaning and dimensions. It is also significant that more and more organizations are now embracing the rhetoric of adaptive capacity and recognizing the need to engage with the processes through which communities adapt. For effective actions, however, it is important that a clear picture emerges so that adaptive capacity can be systematically addressed. If communities are to address even known, short-term challenges, access to knowledge of and the capacity to employ coping or vulnerability reducing measures is essential. As environmental changes bite, the ability of indigenous knowledge to support adaptation will decline and 'the resources and space for adaptation should become a central development imperative' (Adger, Huq et al., 2009: 308). Systems of support that are mandated to provide the most vulnerable with the opportunity to explore and achieve adaptations to their lives and livelihoods need to be seen as essential.

Adapting development

In this light, building adaptive capacity shifts from a component in an adaptation intervention, to the necessary outcome of development actions. Climate change is defining a new direction for development, in which the goal is to provide communities with sustainable access to the resources necessary to meet an uncertain future. Adaptation needs to be seen as 'an opportunity for social reform, for questioning the values that drive inequalities in development and our unsustainable relationship with the environment' (Pelling, 2010: 3). Adaptive capacity demands opportunities for communities to engage in learning cycles, so that they can test and revise alternative ways of living in the face of emerging environmental change. The focus is structural rather than narrowly needs based, technical or economic, drawing attention to the processes that mediate information, knowledge and resources. The manner in which systems of governance and the quality of relationships define access to and leverage in formal and informal institutions is a central concern, insofar as it is through these institutions that adaptation actions will be identified or demanded, planned for, accessed and distributed. Attention is drawn to the determinants of exclusion and marginalization, including culture, gender, ethnicity and, ultimately, power. In this view, development becomes a process of transforming relationships through actions that amplify the voices of the poorest. Power holders are recast as duty-bearers with the responsibility to fulfil rights, enabling the 'development of claims that seek to empower excluded groups and that seek to create socially guaranteed improvements in policy' (Uvin, 2004: 163).

This book offers a viewpoint on how adaptive capacity can be supported through development practice to achieve these ends. Working from a resilience perspective, and using examples and short case studies, the following chapters build a picture of adaptive capacity, and how poor and marginalized communities can best be supported as they deal with the new challenges that climate change will bring. Resilience is used to explore how systems of people and ecosystems can maintain desirable characteristics in the face of climate change. But a focus on social actors pushes adaptation further, to encompass social and political processes that enable a fundamental change in the power relationships that assist learning and, ultimately, define what constitutes a desirable future. This framing reflects a view of adaptation that encompasses both resilience and social transformation (Pelling, 2010) and has much in common with recent thinking on the relationship between adaptive capacity in environmental governance (Plumber and Armitage, 2010). Here, much of the focus is on rural areas, where livelihoods are particularly sensitive to changes in rainfall, temperature, seasonal onset and the frequency of extreme events. A new framing of development in these communities is urgently required.

Chapters 2 and 3 are concerned with the definition and content of adaptive capacity. Chapter 2 is focused on theory rather than practice, seeking to better understand what is meant by adaptive capacity. It does so first by locating it in relation to resilience. This identifies adaptive capacity in terms of how communities are able to increase the prospect of maintaining their well-being. The second half of the chapter highlights how studies of resilience in real systems have revealed the nature of the challenge that adaptive capacity has to overcome. All communities live in complex environments, where complexity makes understanding the impacts of change difficult to predict. Uncertainty is therefore a central feature not only of climate change but of all aspects of life. The function of adaptive capacity, then, is not only to enable people to make changes, but to do so in ways that reduce the likelihood of communities experiencing detrimental surprises.

While chapter 2 provides the theoretical backdrop for the book, chapter 3 moves on to identify the implications of resilience thinking for development practitioners. The intention here is not to offer indicators of adaptive capacity that would allow comparative analysis. Rather, it is to identify the different dimensions of how development actions can support adaptive capacity, enabling the context specific assessment of how adaptive capacity can be strengthened. It also brings a process perspective, in which the capacity to adapt is determined more by relationships, knowledge and learning than easily quantifiable outputs. The cross-cutting importance of power in particular is identified. However, the emphasis is on identifying how power relationships can be reformed in favour of adaptive capacity, with attention on how shared processes that draw in many different stakeholders can build a better understanding of complex environments and the alternative strategies for overcoming climate change impacts. *Power sharing* is therefore the first and overarching dimension of support for adaptive capacity, alongside *knowledge and information* and *experimentation and testing*. Knowledge and information emerge from multi-stakeholder environments, in particular by combining different worldviews that provide alternative perspectives on complex problems. These processes in turn inform and are themselves informed by experimenting and locally testing new approaches to livelihood challenges. Exposure to and testing of alternative practices is essential to developing adaptation strategies, preventing communities from becoming too rigid or fixed in their understanding of the challenges they face. Taken together these three components form a learning system, in which communities gain the power, knowledge and opportunities to form and revise their responses to climate change. The result is a framework of support for adaptive capacity that enables communities to build their resilience.

Chapter 4 presents case studies of development practice to demonstrate how this support can be realized. These examples show that rather than being new territory for practitioners, there is body of experience of working with power sharing, knowledge and information, and experimentation and testing that exists within the development sector. Supporting adaptive capacity is not, then, so much a challenge of breaking new ground in development, or of abandoning traditional development outcomes that respond to the immediate needs of communities. Rather, it is one of adjusting the mindset of practitioners and donors so that the three dimensions become central to *how* interventions achieve these outcomes. Four lessons are drawn from rights-based approaches that demonstrate what this altered perspective means in practice: that working on adaptive capacity demands a form of participation that transforms relationships, is political, identifies the state as accountable, and may invoke the law as a tool to achieve support for poor and marginalized communities. In the final two sections, consensus building approaches and participatory technology development are considered in detail as structured approaches to developing power sharing relationships, fostering knowledge and sources of information, and developing a community's capacities and opportunities for experimentation and testing of alternative livelihood strategies.

Finally, chapter 5 concludes by setting this approach to supporting adaptive capacity in the context of 4°C or more of global warming. Rapid warming is now more likely than not, bringing the prospect of greater future uncertainty, amplified impacts and frequent changes. With this reality comes the prospect of communities having to develop successive incremental livelihood changes in response to emerging climate impacts, underlining the urgency of attending to adaptive capacity. Greater warming also increases the likelihood of livelihoods becoming untenable, resource competition spilling over into conflict, and migration emerging as a significant adaptation response. Empowering communities through support for the components of adaptive capacity identified here provides one way of ensuring that the poor and marginalized become better placed to address these profound challenges.

CHAPTER 2
Understanding adaptive capacity

This chapter introduces resilience and complex systems in order to understand the role played by adaptive capacity in processes of change. A systems perspective is adopted to describe how the people, institutions and ecology that make up a community are connected. It is argued that the concept of resilience – a term already used in many different contexts – can most usefully be applied by development actors as a way of describing a system. For example, resilience would refer to the amount of external disturbance that a small-scale farming community could withstand before the community members lose livelihoods or well-being. Adaptive capacity, on the other hand, determines how people can make changes to their system. It is through adaptive capacity that people are able to change their livelihoods to make them more resilient – which is to say better able to withstand external disturbances such as droughts or shifts in seasonal temperatures. In this view, adaptive capacity encompasses the ability to secure incremental changes to existing livelihoods as well as more profound transformative changes towards new ways of living.

In the second half of the chapter, the nature of complex systems is explored. Complexity makes life difficult to predict, meaning that human systems are prone to breaches of resilience that can lead to a loss of well-being. Examples include the over-exploitation of resources such as fisheries leading to an unexpected collapse in fish stocks, or the failure to recognize and make appropriate changes to mitigate and adapt to climate change. But the study of complex systems also reveals how communities can be supported to enhance their adaptive capacity, and thereby better avoid the life or livelihood threatening impacts of climate change. This framing provides the background for chapter 3, in which the challenge of supporting adaptive capacity is defined in terms of different dimensions of development focus.

Resilience and adaptation

In development, adaptive capacity and resilience have come to mean different things to different people. Resilience in particular can be a problem as many different fields of study have adopted it as a term with precise, and often very different, meanings. The close relationship between resilience and adaptive capacity can make the difference between them difficult to articulate, even though both are appearing with increasing frequency in development discourse. The result is considerable confusion in the language that is used to describe the different ways in which communities interact with their changing social, political, economic and environmental context (Handmer and Dovers, 2009; Smit and Wandel, 2006).

In their comprehensive book on the topic, Gunderson and Holling suggest that there is a major distinction to be made between 'engineering' and 'ecological' interpretations of resilience (Table 2.1). Engineering resilience describes the speed with which a system returns to its normal (equilibrium) state following a disturbance. Here, resilience is similar to the ability to cope with or absorb external disturbances, where the concern is with how quickly a community will 'bounce back' to a stable equilibrium following a shock. Ecological resilience, on the other hand, is principally concerned with the magnitude of disturbance that can be absorbed before a system changes its structure and function (Holling and Gunderson, 2002).

The ecological definition is rooted in the complexity of living systems in which change, rather than equilibrium, is understood to be normal. First developed from the study of ecosystems, it is now routinely applied to coupled social-ecological systems such as those formed by the rural livelihoods of communities in developing countries. It rejects the idea that a system has a single, normal equilibrium, recognizing instead that there may be multiple alternative stable states and that 'regime shifts' between them are possible. For example, disturbances such as nutrient run-off from agriculture, or long-term selective fishing, can shift a lake from a clear water ecosystem to a turbid, muddy environment, dominated by algae. Resilience in this context describes the amount of interference or disturbance that the coupled social-ecological system can withstand while maintaining its form – in this case, remaining as a useful clear water ecosystem. Walker and Salt summarize that, in this view: 'a resilient social-ecological system has a greater capacity to avoid unwelcome surprises (regime shifts) in the face of external disturbances, and so has a greater capacity to continue to provide us with the goods and services that support our quality of life' (2006: 37).

'Social resilience' places a greater emphasis on the human actors in the system (Adger, 2000). Tompkins and Adger's useful definition refers to social resilience as the 'ability of groups or communities to adapt in the face of external social, political, or environmental stresses or disturbances' (2004: 5). This shift in focus explicitly links resilience to adaptation and highlights the fact that human actions can dominate in social-ecological systems (Walker et al., 2004). This definition also makes clear that the actors in a system can make changes, enabling it to retain the same overall function or identity in the face of disturbances. An example here is where a different combination of crops is

Table 2.1 Summary of resilience definitions

Resilience definition	Characteristics
Engineering	Speed of 'bounce-back' to a natural equilibrium condition.
Ecological	Degree of disturbance before the system changes. Change is normal. Alternative stable states rather than single equilibrium.
Social	As for ecological resilience, but emphasises the ability of people to adapt a system in order to maintain its desirable characteristics.

employed to retain the food producing function of an agro-ecosystem in the face of resource depletion or altered weather patterns. Table 2.1 summarizes these three approaches to resilience.

Defining adaptive capacity

In ecological and social resilience, adaptive capacity and the ability to absorb shocks and disturbances both contribute to the ability to maintain a particular system function. As such, they are combined as different attributes that make up resilience. However, from the perspective of development practice, this presents difficulties for trying to understand adaptive capacity. Adaptation is focused on how the actors in a system can effect change. As a function of adaptive and absorbing capacities, resilience entangles the component of interest – adaptive capacity – in favour of characterizing the overall system. But, as discussed in chapter 1, the practical implications of a call to absorb disturbances are much better understood by development actors than one that focuses on the ability to learn and adapt. Combining absorbing and adapting in resilience blurs a distinction that for development practitioners needs to be emphasized and made clear.

A further source of confusion arises from how resilience is defined. Resilience refers to the ability of a system to retain its overall function, suggesting that, in turn, adaptive capacity is concerned only with enabling changes that ultimately maintain the existing system. This ignores the possibility of actors choosing to effect more profound, transformative, change. Rather than allowing adaptive capacity to encompass actions that result in significant livelihood changes, 'transformability' is introduced as a separate characteristic that sits alongside resilience and is different to adaptive capacity (Walker et al., 2004; Nelson et al., 2007).

Treating resilience and transformability separately is unhelpful for actors in the development sector who understand their adaptation work to be supporting communities to meet the challenges of climate change. As governments continue to fail to set and reach emissions targets, the prospect of profound climate change becomes increasingly real and those working with communities facing climate change must be ready to address livelihood transformations. If adaptation support is limited to shoring up existing livelihoods then it is bound to fail – and offer false hope to – the communities whose livelihoods ultimately become untenable under increasingly severe climate change. To be meaningful in practice, adaptive capacity needs to be defined in such a way that it enables communities to engage with all viable future options, including both the modification of existing livelihoods and the transformation into new ones.

Adopting an adaptation perspective helps to bridge the development and resilience perspectives. By shifting the focus onto processes of change, adaptive capacity and absorbing capacity can be seen to play very different roles. Figure 2.1 illustrates the alternative processes that disturbances can stimulate, with purposeful actions that employ adaptive capacity in the top part, and passive responses that rely on a pre-existing capacity to absorb disturbances in the lower part. The fundamental functions of the system may be maintained following a disturbance, or alternatively, the system may be transformed purposefully by the actors within it or passively as a consequence of external disturbances.

The different pathways in Figure 2.1 illustrate the crucial role that actors within the system can play in *enhancing* resilience, if they are able to make purposeful changes in response to actual or predicted disturbances. Also acknowledged is the fact that many social-ecological systems are vulnerable to particular disturbances, meaning that they emerge with the ability to carry out the same functions, but with a degraded ability to persist in the face of future disturbances – that is, they emerge with lower resilience (Chapin et al., 2006). Ideally, however, disturbances stimulate adaptive capacity, yielding actions through which the actors make incremental or more profound changes to their social-ecological system, increasing their resilience, transforming the system to a completely new regime, or reducing vulnerability to particular shocks or stresses.

In this view, adaptive capacity is linked to transformation as much as it is to resilience, even though they are defined in mutually exclusive terms:

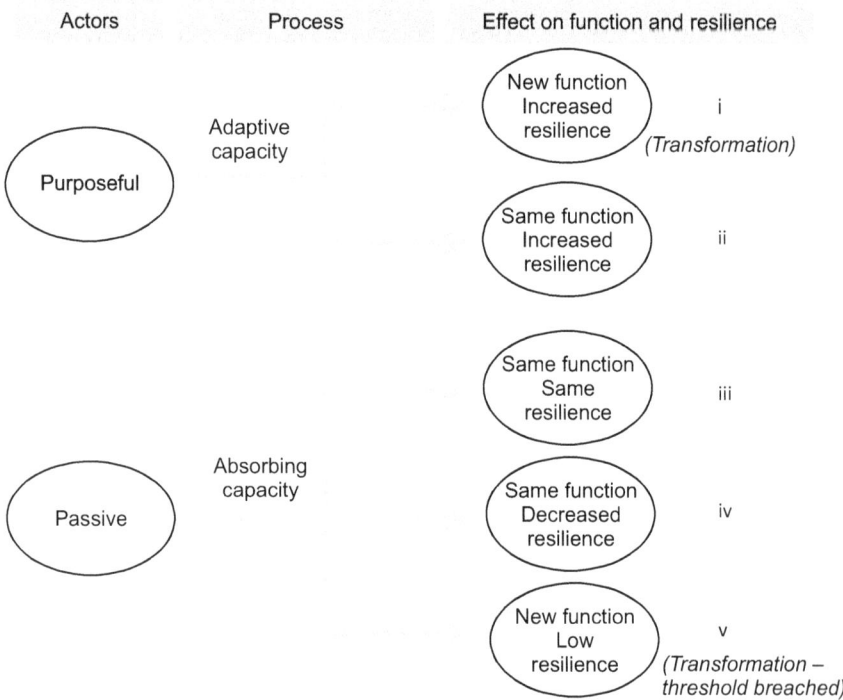

Figure 2.1 An adaptation perspective on changes to a social-ecological system following a disturbance

in resilience the system maintains its current function, in transformation an entirely new stable regime is achieved. However, from a practical perspective, this distinction breaks down. Many of the attributes of resilience and transformation are the same: the knowledge and resources to make changes to livelihoods are necessary whether the outcome is the maintenance of resilience in the face of new threats, or the transformation of livelihoods to capitalize on changing circumstances (Walker et al., 2004). Armed with knowledge of emerging climate change, communities in some contexts may choose to adopt new ways of earning a living (or places in which to live) rather than attempt to sustain existing livelihoods.

Many with cattle ranching livelihoods, for example, converted their holdings to attract ecotourism following trade, drought frequency and ecosystem changes in south-eastern Zimbabwe in the 1980s (an example of pathway (i) in Figure 2.1; Walker et al., 2004). Here, transformation involves *overcoming* the resilience of the existing system through the mobilization of adaptive capacity. Yet for other actors in other circumstances, adaptive capacity may be utilized to build resilience through the awareness of and capacity to employ opportunities for incremental changes to livelihoods that enable communities to cope with or avoid the impacts of climate change. For example, having secured access to climate-related information, pastoralists in Niger opted to proactively sell livestock prior to a drought, thus yielding the resources necessary to cope. In this example, the ability to cope and recover was enhanced through the community's capacity to engage with and respond to the prospect of drought; that is, through their adaptive capacity (an example of pathway (ii) in Figure 2.1; Ensor and Berger, 2009a: 115-130). The remaining pathways in Figure 2.1 can be illustrated, albeit simplistically, by pastoralist livelihoods in the face of changing drought frequencies. While regular drought can be coped with through the loss and rebuilding of cattle numbers (pathway (iii)), increases in drought length and frequency gradually reduce the capacity to rebound as livestock numbers have insufficient time to recover (iv), ultimately leading to the loss of livelihood should no external support be available (v).

Definitional distinctions, then, are important for identifying where development effort needs to be placed, but should not be confused with the existence of neat boundaries on the ground. Community-based adaptation is understood in terms of empowering communities to make changes to their lives and livelihoods, through increasing resilience or reducing vulnerability of existing livelihoods, and through access to the opportunities to make transformations in their lives. The goal for development actors is to build adaptive capacity in communities so that these different actions can be taken in context, with adequate support, and when necessary. The focus is on the potential for actors within a particular human and environmental system to respond to changes, shape changes, and create changes in that system (Chapin et al., 2006: 16641).

Resilience thinking

The definition of resilience discussed above derives from extensive studies of how ecological and coupled socio-ecological systems behave. Inspired by Holling's seminal 1973 paper 'Resilience and the stability of ecological systems', this body of work has enhanced understanding of how these systems behave and, in turn, why efforts to manage ecosystems often fail. The insights of several decades of research are captured in Walker and Salt's 2006 book *Resilience Thinking*, a title that is borrowed here for the attention it draws to the mindset change that Holling's resilience framing stimulates. This mindset is significant for understanding adaptive capacity and how the challenge of adaptation can be understood and addressed. In particular, three contributions of resilience thinking set the scene for thinking about adaptive capacity: the nature of complex systems, the significance of thresholds, and the four stages of adaptive cycles.

Complexity

All communities or organizations are situated within ecosystems and economies that in some sense value those that perform well. This creates selective forces that influence the overall structure of the community: it is a self-organizing (or 'adaptive') system, changing in response to external influences and as a result of internal relationships. The self-organizing and interactive nature of the component parts (people, plants or animals) make life unpredictable, even when the separate elements are well understood (Holling and Gunderson, 2002). Yet these 'complex adaptive systems' are the norm in human and ecological contexts, occurring whenever there are (Walker and Salt, 2006: 35):

- components that are independent and interacting;
- selection processes at work on the components; and
- constant occurrences of variation and novelty through component changes or additions.

Unpredictability means that surprises are to be expected. The actions of people or parts of the system that are remote in time, space or scale are mediated through the interactions in complex systems, resulting in unexpected consequences. Human actors that underestimate complexity or assume linear cause-and-effect responses are particularly likely to encounter surprises. Moreover, the more that people rely on familiarity with only a small part of complex systems – their local area or their lifetime of experiences – the more their assumptions about the world become increasingly vulnerable to surprises.

In 2003, a paper by Carl Folke described this problem in terms of our relationship to natural resources. The way of thinking that has come to dominate in natural resource policymaking assumes that the capacity of the biosphere to support human development is unlimited: we can relentlessly exploit the goods and services that ecosystems provide and control environments to optimize their productivity. The implicit assumption is that the environment is infinitely stable and resilient. Resilience thinking tells us that this is not the case. Rather, change and surprise are normal, and our actions determine whether our ability to respond is eroded or enhanced (2003: 2028): 'The concept of resilience shifts perspective from the aspiration to control change in systems assumed to be stable, to sustain and enhance the capacity of social-ecological systems to cope with, adapt to and shape change and learn to live with uncertainty and surprise'

Complexity explodes the myth that natural variability can be effectively controlled and the consequences predicted. There is no 'natural equilibrium' to which we can expect a system to return. This is a reality that has long been appreciated in many indigenous knowledge systems, which recognize that significant but infrequent disturbances are a feature of living within ecosystems (Berkes and Folke, 2002: 146). The contrast between the assumptions in 'command and control' and complex system management is summarized in Table 2.2, highlighting the nature of the challenge that an appreciation of complexity uncovers. This side-by-side comparison also draws attention to the extent to which thinking needs to change in those whose training and experience have been dominated by assumptions of linearity and controllability.

From an adaptation perspective, an appreciation of complexity means an appreciation and acceptance of unpredictable change. It requires a mindset shift away from an assumption that an understanding of the current ecological and social context is sufficient, and towards a focus on the capacity to adapt to changes yet to be experienced. From a complexity perspective, uncertainty is significant to adaptation not just a result of our limited understanding of climate science (chapter 1), but also as it is inherent in socio-ecological contexts – in the life of communities – due to the capacity for surprises that resides in complex systems.

Thresholds and slow variables

The reality of changes and surprise extend to the potential for social-ecological systems to 'flip' from one form to another. 'Function', 'regime' or 'operating state' changes, such as from clear to turbid lakes, or grassland to shrub-deserts,

Table 2.2 Shift in thinking and perspective for complex systems

Command and control	Complexity and resilience
Assume stability: control change	Accept change: manage for resilience
Predictability: optimal control	Uncertainty: risk spreading and insurance
Managing for increased yield	Managing diversity for coping with change
Technological change solves resource issues	Management systems designed to build resilience

Source: adapted from Folke, 2003: 2033

are surprising or unavoidable precisely because ecosystems are complex. The feedbacks and connectivity mean that, in these two examples, nutrient run-off can set in train an irreversible decline in water quality, or overstocking can reduce grasses to levels that prevent the periodic grass fires that control woody growth. These fundamental changes – the flipping of the system function – occur when a threshold is breached, after which the regulating feedbacks (water oxygen level or periodic fire) change, holding the system in a new state.

Resilience refers to the amount of disturbance that a system can withstand before crossing a threshold into a new state. How much nutrient pollution a lake can withstand before flipping into turbidity defines its resilience, but also changes over time and response to other system characteristics. The speed of flow in tributaries or the prevailing temperature of the water may shift the degree of nutrient discharge that can be withstood. Critically, however, the accumulation of nutrients in the lake sediment will change the lake's resilience. A large run-off event may render the water temporarily murky but – so long as no further nutrients are added – the lake will recover as the nutrients are absorbed into the sediment. However, even a small amount of further nutrient addition at this stage can render the lake unable to recover.

The fate of lake systems highlights the importance of the background changes in the system. Ecosystem collapse only occurred due to the slow accumulation of nutrients in the lake sediment. Slow variables – those aspects of the system that change over long timescales – are particularly significant because they can have a dominant effect on the resilience of the overall system. They shift thresholds that other, faster changing dynamics depend on. This is a key dimension of climate change, which brings slow and often imperceptible changes in weather patterns that can dramatically impact on the ability of social-ecological systems to recover from disturbances. Temperature increases that change the prevalence of disease vectors or the quality of foliage, for example, can reduce the health of cattle rendering them susceptible to periods of drought that they would have otherwise have survived. Slow variables also demand particular attention because they can be hard to spot when the focus is on shorter-term challenges or changes (see, for example, ill-fated attempts to research the relationship between insects with short lifetimes and centennial forest regeneration: Holling and Gunderson, 2002: 30). In complex social-ecological systems, the possibility of overlooking the effect of climate change on key thresholds is very real – and the more so in the face of the short-term challenges that characterize poor communities.

Adaptive cycles

Complexity and thresholds are drawn together in a visual metaphor known as the adaptive cycle (Figure 2.2). Its purpose is not to be a predictive model for the lifecycle of complex systems, but to help break down and understand their dynamic nature by drawing attention to four distinct phases: rapid growth, conservation, release and reorganization. Since its introduction, researchers

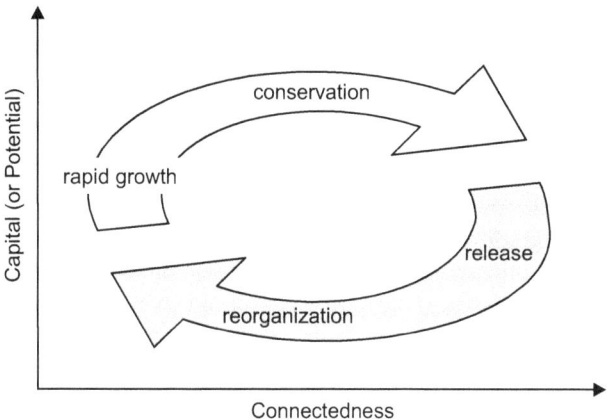

Figure 2.2 The adaptive cycle
Source: Walker and Salt, 2006: 82

have found that most social-ecological systems, human organizations and economies follow these recurring cycles, with the connections and feedbacks different in each phase (Holling and Gunderson, 2002: 33). Essentially, it describes short periods of rapid, sometimes catastrophic change (release) followed by longer periods of development and consolidation.

The four phases are associated with changes in capital and connectedness, shown on the two axes in Figure 2.2. In this context, capital refers to the accumulated usable knowledge, innovations and skills, or the quality of networks of relationships, or, in an ecosystem, the amount of biomass and nutrients. An accumulation in capital occurs as societies or systems grow in structure and efficiency, and provides the potential for other types of futures to emerge: networks of trust and reciprocity built in one setting, for example, can be deployed in another. As a result, this dimension of the adaptive cycle is also sometimes referred to as 'potential'. Connectedness refers to the internal connections and feedbacks in the system, so that high connectivity results in a system that regulates itself, insulated from external influences. Importantly, resilience also alters through the cycle, and can be added to the image as a third dimension. As Holling and Gunderson explain, these three properties shape how a system changes over time (2002: 51):

> Potential [capital] sets the limits for what is possible – it determines the number of alternative options for the future. Connectedness determines the degree to which a system can control its own destiny, as distinct from being caught by the whims of external variability. Resilience determines how vulnerable the system is to unexpected disturbances and surprises that can exceed or break that control.

As Figure 2.2 illustrates, the adaptive cycle is broken into two modes, known as the fore loop (rapid growth and conservation) and the back loop (release and reorganization). The two loops have very different characteristics. The fore loop is associated with increasing stability, growth and the emergence of a settled society, such as in the succession of trees into an established forest, or where a new business moves from a period of innovation into a mature organization with an established market share. Rapid growth occurs as species or people exploit opportunities or resources, such as newly available nutrients, space and light in a recently cleared forest setting. As the weak connections between scattered or newly formed components gradually coalesce, the system moves into the conservation phase. Here, efficiency and specialization become the most rewarded attributes, favoured over experimentation and opportunism that constitute success in the rapid growth phase. Those species, people or organizations that succeed lock up capital (biomass in large trees, or assets, skills and experience in large organizations) and become increasingly internally connected, losing flexibility in favour of efficiency in resource use. This is a slow process, but ultimately resilience decreases and a form of brittle rigidity sets in, with systems becoming fixed and very stable, but over a limited set of conditions (Walker and Salt, 2006: 76–82).

The back loop is very different, characterized by uncertainty, novelty and experimentation. It starts when the system's resilience is exceeded, breaking apart connections and unlocking stored capital. The release phase is short and chaotic, rapidly making available the raw materials of a new cycle in a wave of 'creative destruction' (Schumpeter, 1950). As Holling and Gunderson put it, 'the trigger might be entirely random ... such events previously would cause scarcely a ripple, but now the structural vulnerability provokes crisis and transformation because resilience is so low'. For a short period, feedbacks become destabilizing, as the insect defoliator or annoyed shareholders make merry with the forest trees or organizational bureaucracy. The process burns itself out as the insects run out of food or the firm fires workers and managers, setting the stage for restructuring (2002: 45). With the breaking apart of the established order, the complex system enters a period of uncertainty and unpredictability.

Reorganization occurs as invention and experimentation yield novel ways to exploit the released capital – pioneer species or entrepreneurs provide novelty and diversity in place of relative order of the fore loop periods of growth and conservation. Reorganization is a transition in which the chaos and disorder of release gradually gives out to new constraints as the actors in the system rearrange themselves around newly identified opportunities. Three alternative scenarios can emerge – a repeat of the previous cycle, reorganization and growth in a novel form, or a maladaptive, degenerative state in which low connectedness, low capital and low resilience form a vicious cycle, referred to as a the 'poverty trap' (Holling et al., 2002: 96). Communities that, following a trauma, lose social coherence, assets and adaptive capabilities, leaving their members unable to re-establish sufficient capital to stimulate growth, provide one example of this phenomenon.

Finally, in resilience thinking 'panarchy' refers to the presence of multiple, connected adaptive cycles that operate within the same system but at different speeds and geographical scales. Work on panarchies has highlighted that while the scale of focus for communities is usually local, their ability to respond to disturbances will depend on interactions with the scales above and below. For example, rules and regulations that (usually) change slowly and cover a wide area may limit the ability of individual farmers to develop irrigation systems or pastoralists to access grazing lands, while support during prolonged drought events may depend on the political and economic conditions at the scale of the national government or international aid agencies. On the other hand, successful experiments in local resource management, for example, can also percolate up to higher scales, influencing local actors or changes to national or regional policy. These two cross-scale linkages – the regulating or restraining of behaviour imposed on faster cycles by slower ones, and the innovating and creative impulse that faster scales transmit to slower ones – are typical in most systems and help to resolve the tension between growth and conservation. As Holling et al.'s description suggests, it is a characteristic that contributes to the persistence of the overall system: 'The fast levels invent, experiment and test; the slower levels stabilise and conserve the accumulated memory of past successful, surviving experiments. The whole panarchy is both creative and conserving. The interactions between cycles in a panarchy combine learning with continuity.' (2002: 76)

Resilience thinking for adaptive capacity

Resilience thinking tells us that human systems are complex. Change occurs in ways that are difficult to predict. Even small disturbances can upset an established way of life, leading to the emergence of new winners and losers in a system that performs a very different function. The adaptive cycle identifies the challenge that complexity brings: how to grow the well-being and stability that emerge in the fore loop, without risking the catastrophic losses of release during the back loop. How can novelty and diversity be maintained when the selective pressure for efficiency increases as a system matures? How, for example, can a focus on optimizing yields (and profits) in a farm system be tempered by crop diversity and experimentation? By employing foresight and intentionality, human actors can reduce or eliminate the 'boom and bust' character of some adaptive cycles (Holling et al., 2002).

If adaptation is about avoiding catastrophic change and poverty traps, then its task is to identify the slow variables that bring thresholds closer, and do one of two things: to move a social-ecological system back from conservation to growth – releasing capital and connectedness, reducing rigidity, and increasing the resilience of the existing system; or to stimulate a controlled transition to reorganization and growth with the minimum possible cost. Barriers to effective adaptation actions, including inequitable or inflexible values or governance arrangements, can lock communities into livelihood practices that

24 UNCERTAIN FUTURES

rely on absorbing capacity and may ultimately lead to unwelcome surprises (Adger, Lorenzoni et al., 2009).

It is significant that in many cases local, indigenous knowledge recognizes the necessity of coexisting with change and acknowledges complexity, trading efficiency or optimal productivity in favour of living with uncertainty. This understanding means disturbances are allowed to induce small-scale changes to livelihoods, in the knowledge that the alternative is to remain inflexible to the point that accumulated disturbances will bring about catastrophic release (Folke et al., 2005). For example, traditional controls over stream fishing rights by the indigenous peoples of the American Pacific coast were sensitive to changes in salmon populations induced by ocean conditions, reducing catches where necessary but ensuring the long-term sustainability of the fishery (Trosper, 2003). Pastoralism accepts the costs of transhumance in return for the ability to live with harsh conditions and periodic drought. Looking to the future, Walker and Salt talk in terms of the alternatives facing Europe's energy mix in the face of 'peak oil' and an inevitable decline in fossil fuel availability: there can be either a 'reverse move, from late conservation back to early conservation through small-scale changes' that provide renewable energy options, or 'a direct move at the scale of the whole industry through a rapid renewal phase to a new front loop' (2006: 85). To not plan for either strategy commits the industry, and well-being in Europe, to a period of uncontrolled collapse. Figure 2.3 illustrates these possibilities.

Climate change and its impacts have emerged as key slow variables of concern in development practice (although they are by no means the only such variables). Yet, as discussed in relation to Figure 2.1, adaptive capacity is significant as it represents the ability of social actors to make deliberate changes. In so doing, they can achieve either of the options illustrated in Figure 2.3, thereby preventing the slow variable from reducing a threshold to the point that an uncontrolled release – and loss of human well-being – occurs. Human actors can induce incremental change to enhance the resilience of the existing system (moving from conservation back towards rapid growth) or transformational change to change the function of the system altogether (jumping

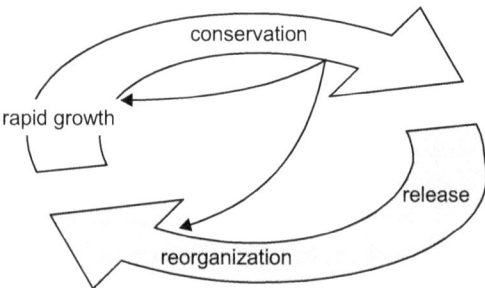

Figure 2.3 The adaptive cycle and options for adaptation
Source: adapted from Walker and Salt, 2006: 83

from conservation to reorganization and rapid growth). Actors implement these alternative strategies by employing adaptive capacity, as illustrated by pathways (i) and (ii) in Figure 2.1. If absorbing capacity is relied on in place of adaptive capacity, the position in the fore loop may be unchanged following a disturbance (pathway (iii) in Figure 2.1), or the threshold brought closer (that is, a loss in resilience, as in pathway (iv)), or the threshold crossed, precipitating a shift into the release phase of the back loop (pathway (v)).

The idea of 'panarchy', or the presence of adaptive cycles that operate at different scales, provides insight into how adaptive capacity enables actors to secure the alternative strategies illustrated in Figure 2.3. The lessons from small or fast-scale experimentation and testing can be accumulated as learning at the slower scale, enabling stability and continuity to be integrated with novelty and change. As noted above, successful local experiments can also percolate up to larger scales, influencing the behaviour of actors or changing policy. However, the ability to adapt is also defined by the prevailing social and political context, determining how assets and information are distributed and who decides how they are used. While some elements here are local (such as networks of family relationships), it is also important to recognize that broader and sometimes global social, economic and political forces may have the most significant influence at the local level, as where international free trade agreements remove supportive subsidies or price guarantees for a particular local crop (Smit and Wandel, 2006). It may not be sufficient to consider only micro-scale relationships if it is 'powerful political and economic vested interests that determine the nature of the adaptation context' (Brooks, 2003: 12).

Looking beyond the community scale is, then, essential in understanding local adaptive capacity. Adaptation options can emerge from small-scale experiments with alternative livelihood practices that, if adopted at the local scale, can increase resilience, shifting communities back down the fore loop. The formal and informal institutions that surround local communities, at different proximities, play a key role in mediating interests across scales, shaping the opportunities and constraints for local-level changes. In poor or marginalized communities where the continuity and resilience of inequitable or unsustainable resource access and distribution arrangements is undesirable, the capacity to influence and create change at the higher scale can be critical. Creating the conditions for effective participation through which local knowledge can gain traction in the bureaucracies at higher scales is thus a key challenge for local adaptive capacity (Jennings, 2009).

Dealing with complexity

The need to support relationship building across scales is recognized by the Swedish Commission on Climate Change and Development. The Commission identify governance as a pillar of adaptive capacity, finding that 'the fullness of the institutional environment provides the means through which people, working with others, have access to resources, articulate needs, and exercise their rights' (2009: 4). Adger similarly focuses attention on how 'networking capital' between local communities and those with shared interests in governing elites can be the foundation for a productive and democratic relationship to the state (2003). Ideally, the presence of such networks facilitates the two-way flow of information: firstly, upwards from household and community to improve policy responsiveness to the local socio-economic and environmental context; and secondly, downwards in the dissemination of knowledge (relevant meteorological science, adaptation options) and resources. The experiences relayed by Ensor and Berger (2009a) illustrate these processes in action. In Kenya, the integration of meteorological office staff into an adaptation project team ensured forecasting information was generated for the project area and converted into a format useful for local farmers (for example, appropriate crop varieties for the anticipated rainfall), while in Sudan strong links between a water stressed community and state government enabled support for adaptation activities through improved information, and is helping the community to secure a space for their views on the construction of a new dam.

Fermenting cross-scale interactions has been identified as critical in many contexts for supporting effective resource management and in addressing environmental change at the local level (Cash et al., 2006). These relationships are most commonly built in multi-stakeholder fora that bring together divergent interest groups for a limited time to explore resource allocation or adaptation issues (Nelson et al., 2007). Osbahr et al. describe one such initiative in Mozambique, in which the potential to enhance resilience and protect spaces for livelihood transformation was generated because, 'government, NGOs and local communities have helped to create emergent conditions for local-level adaptation to drought, food insecurity and poverty reduction. These deliberately operate across traditional levels of authority for the management institutions' (2008: 1962).

The success of cross-scale approaches is due to the potential of those with different perspectives, when working together, to rethink the management of vulnerable systems. This is achieved when the approaches (Bunce et al., 2010):

- emphasize the need to understand the properties of a linked social-ecological system, the stressors and drivers of change which impact upon it and the interactions and feedbacks between them (that is, complexity);
- enable strategies where stakeholders are actively engaged in planning and implementing policies;
- explicitly look at the cross-scale linkages and impacts of change, and develop multi-level institutions to make sure that the local and national, regional and even global priorities are brought together.

'Social learning' is used to refer to processes that enhance common knowledge, awareness and skills by engaging multiple participants, sharing diverse perspectives, and thinking and acting together (Tschakert, 2007). Folke et al. refers to adaptive co-management in this context, defined as a 'flexible community-based system of resource management tailored to specific places and situations, supported by and working with various organisations at different levels' (2005: 448). Flexibility implies that the approach is structured to allow for learning and to generate ways to respond to and create change. In this way, adaptive co-management allows adaptive actions to be tested, monitored and re-evaluated as necessary, but also goes further by emphasizing the need for a form of problem solving that occurs across traditional organizational levels. Tompkins and Adger similarly talk of building resilience through expanded spaces of engagement and flexible learning-based management (2004). Social capital is a necessary component, built through investment in relationships and networks and embodied in trust between collaborators. It increases flexibility and provides the basis for shared problem solving in both formal and informal settings. While informal environments can be more innovative and flexible than formal bureaucracies, and often are the forerunner to formal systems, they share a reliance on social capital in order to provide an environment for social learning (Folke et al., 2005).

Learning, then, can prevent responses to climate change becoming tied to a particular set of actions, and thus provides a mechanism to deal with uncertainty and change. By marrying this with linkages and collaboration across scales and actors, policy and management can also become more responsive to the needs of diverse stakeholders, 'particularly of local groups who have been marginal to the decision-making processes impacting on their lives' (Bunce et al., 2010: 495). Indeed, the benefits of an approach predicated on inclusivity extend to the giving of voice to marginalized stakeholders, the integration and recognition of diverse knowledge systems, and increases in the depth of civil society and citizenship (Nelson et al., 2007). However, the purpose is not only to bring together different interests in decision making, but to ensure that the adaptive capacity that results includes the ability to adjust to changes that are occurring at multiple scales (Chapin et al., 2006).

Collins and Ison (2009) highlight the contrast between the social learning paradigm and conventional, command and control policy responses that focus on behaviour change through regulating, informing or educating actors. By recognizing that adaptation is addressing complex problems, and therefore that each stakeholder's knowledge of the situation is incomplete, it becomes clear that adaptation requires not only a change in practices but also a change in understanding of the problem. Adaptive behaviour is, therefore, defined by these changes and emerges from social learning processes. Collins and Ison summarize that (2009: 370):

> this perspective recognizes that adaptation requires learning (understood as doing) rather than simply participation, and specifically a form of

learning which is collective in nature. The term social learning has arisen in response to a growing recognition that learning occurs through situated and collective engagement with others.

The alternative – the absence of flexible, inclusive cross-scale learning – characterizes what Brown has referred to as institutional 'misfits': policy failures that are due to a mismatch between the scales of decision making and the social ecological system being managed (2003). Bunce et al. describe a misfit in coastal development policy in Tanzania and Mozambique, where large-scale interventions made local communities more vulnerable. Here, benefit was sought for society at large, through, for example, the exploitation of 'underutilized' resources; however problems arise as (2010: 494):

> firstly, regional and national development priorities may not take local impacts and local vulnerabilities into account. Secondly, these developments may not take other stressors and particularly the impacts of climate change into account. These two factors, combined, mean that externally driven interventions may actually undermine resilience.

Tompkins and Adger underline this sentiment, summarizing that (2004: 10):

> Adaptation to both gradual and significant changes should involve encouraging the evolution of new institutions that are sensitive to the resilience of the ecosystems they are managing and knowledgeable about the specific nature of the risks of climate change.

Formal and informal institutions that embrace social learning can provide this link to the specific, local scale. They bring a new meaning to participation, drawing communities into processes that secure a place for their interests and perspective in experimenting, learning and decision making. This is at the heart of adaptive capacity, as it is through effective relationships with others – including government agencies, scientific bodies and NGOs – that the capacity to understand and act on complex problems can be realized. By working across scales, social learning brings to adaptive capacity the ability to collapse panarchies in ways that provide the most vulnerable with opportunities to secure ongoing support that responds to their needs. As such, building adaptive capacity emerges not simply as an issue for climate change adaptation: it is at the centre of achieving sustainable development in an uncertain world.

Key messages from resilience thinking

The purpose of the remainder of this book is to help explore how opportunities to create and support incremental and transformative change can be fostered through development practice. Building on the understanding of adaptive capacity presented here, chapter 3 presents adaptive capacity in terms of three principal components, while chapter 4 moves on to consider how they can be better integrated into development practice. Throughout, the key findings of this chapter that are relied on are that:

- social-ecological systems are complex, unpredictable and linked across multiple scales;
- slowly changing variables, such as climate change, frequently shift the threshold beyond which the system loses its ability to fulfil the function desired by the actors within it;
- adaptive capacity is employed by actors to make livelihood changes that increase resilience, reducing the chance of the system losing its ability to provide its desirable function, or transforming the system altogether;
- the opportunities for livelihood changes may be limited, constrained or regulated by the behaviour of actors at larger scales;
- cycles of experimenting and testing provide the learning to respond to uncertainty and inform livelihood changes;
- the perspective of actors at different scales are necessary to build an understanding of complex systems and to respond to changes occurring at those scales.

CHAPTER 3
Unpacking adaptive capacity

In chapter 2, ideas from resilience thinking were introduced to help understand why a development focus on adaptive capacity is needed to support communities. Future uncertainty was shown to be the reality in a complex world, yet insights from several decades of resilience research were used to provide a way of understanding the challenge of sustaining the well-being of communities. Taking this as its starting point, this chapter moves to the next question – what, then, are the components of adaptive capacity that development actors can support? The answer is provided in terms of three dimensions that provide a framework for those seeking to support adaptive capacity. The remainder of the chapter is then spent unpacking this framework, before chapter 4 moves on to explore it in terms of examples from development practice.

Supporting processes of change

This chapter proposes that supporting a community's adaptive capacity should be seen as expanding their opportunities for change in ways that explicitly address complexity and uncertainty. The focus is on processes of change, through which the enhanced resilience or transformation of livelihoods can be achieved. In turn, the character of formal and informal institutions and networks of relationships that determine the nature of achievable change become significant. The limitations of climate projections, compounded by the complexity of social-ecological systems, mean that the challenge of supporting communities' adaptive capacity is one of identifying the mechanisms through which a better informed understanding of current problems and future scenarios can be generated, assessed and acted on by the poorest in their own best interest.

This understanding contrasts with those found in many other discussions of adaptive capacity in an important way. Routinely, sources also draw attention to the livelihood context, reflecting a view that the determinants of adaptive capacity include tangible assets, such as financial and natural resources, as well as the process components that are the focus here. For example, the Overseas Development Institute's adaptive capacity framework refers to the availability of key assets, including natural, physical and financial capital (Jones et al., 2010), while International Union for Conservation of Nature (IUCN) include stocks and diversity of natural, physical and financial assets in their characteristics of adaptive capacity (Marshall et al., 2009). Processes do remain important in these sources, as both the diversity and distribution of the tangible components are taken to be important. Chapin et al. identify with this perspective, suggesting that adaptive capacity, 'depends on the amount and diversity of social, economic, physical, and natural capital and on the social networks, institutions, and entitlements that govern how this capital is distributed and used' (2006: 16641).

In their examination of adaptation, the Swedish Commission on Climate Change and Development go further, casting adaptive capacity within a broader human development framework. In their view, adaptive capacity increases with human development, defined in terms of four headings – wealth, health, education and governance (Box 3.1). By making the link to human development explicit, this approach makes it clear that the task of supporting communities facing climate change is one that cannot be easily separated from the drivers of poverty and existing vulnerabilities that are the target of development actions. It also identifies adaptive capacity as a context specific challenge that must account for and support local development needs in tandem with building access to knowledge and adaptation outcomes.

It is not the intention of this chapter to argue with this: on the contrary, it is an unavoidable fact that adaptation actions will ultimately be defined by the livelihood context. However, there is a critical difference in terms of how adaptive capacity's relationship with development is perceived. The overall message of this book is to suggest that a reorientation of development is necessary to ensure support for adaptive capacity. To do this does not imply abandoning attention to vulnerability and poverty, but rather requires a focus on the processes through which communities are able to engage in creating and shaping change. Human development underpins communities' ability to be effective in these engagements, but human development on its own will not be enough – it is *how* development proceeds that adaptive capacity draws attention to, rather than the more familiar development question of *what* outcomes are necessary. By thinking of adaptive capacity in this way,

---

Box 3.1 Adaptive capacity and human development

Wealth, or access to assets, provides the buffers and backup that take people through crises and enable them to recover. Assets may be financial or material, directly accessible or through insurance, and come from the social networks of family and kin or through government social protection schemes for those with few means of their own.

Health safeguards the productive capacity of the individual and the integrity of families. This comes through clean water and effective sanitation, safe childbirth, and food of the right kind and amount so that children grow to their full potential.

Education gives people access to information, knowledge of their options, and the ability to make informed choices.

Governance, or rather the fullness of the institutional environment, provides the means through which people, working with others, have access to resources, articulate needs, and exercise their rights.

Source: Swedish Commission on Climate Change and Development, 2009

outcomes arise as a consequence of the processes that a focus on adaptive capacity provides. In particular, the underlying climate change vulnerabilities and risks that a community faces are addressed through the adaptation measures that result from these processes. This understanding suggests that supporting adaptive capacity may require a long-term approach with multiple phases of engagement, using processes to gradually build both access to adaptation measures and the necessary assets and resources to implement those measures.

Sen's capabilities construct is helpful in exploring this distinction. Capabilities refer to the freedom or potential to achieve desirable outcomes: the focus is on whether or not there is the opportunity to achieve a particular outcome, rather than the outcome itself (Sen, 1999). Capabilities can thus be understood as the 'opportunity set' available at a point in time (Gore, 1997), encompassing both assets and the institutional and structural framework. Work supporting adaptive capacity can thus be understood in terms of increasing the available opportunity set in support of achieving tangible outcomes that enhance resilience to climate change. The purpose of the framing presented here, however, is to recognize that adaptive capacity means acknowledging and responding to uncertainty and complexity. The focus is thus on expanding capabilities in ways that allow communities to engage with the challenges of complex systems. This framing explicitly recognizes that adaptation outcomes will ultimately be limited or defined by the existing assets and vulnerabilities – adaptation is always situated in its local context, be it in the developed or developing world. The role of a focus on adaptive capacity is to expand capabilities in directions that enhance the ability to identify appropriate adaptation actions in that context. The required reorientation of development thinking is in identifying adaptive capacity as access to the processes *through which* concrete adaptations can ultimately be secured.

Three dimensions of support for adaptive capacity

With this process focus in mind, the aim now is to identify areas in which development effort needs to be placed if adaptive capacity is to be supported. To understand how people experience and create change, chapter 2 adopted a resilience perspective. In this view, people are seen as part of a system, in which the effects of external influences emerge as a result of the relationships and linkages between people and their environment. The conclusion of chapter 2 suggested that these social-ecological systems have a number of properties of significance for understanding the challenge of adapting to climate change:

- social-ecological systems are complex, unpredictable and linked across multiple scales;
- slowly changing variables, such as climate change, shift the threshold beyond which the system loses its ability to fulfil the function desired by the actors within it;

- adaptive capacity is employed by actors to make changes that increase resilience (reducing the chance of the system losing its ability to provide its desirable function) or transform the system altogether.

Building from this understanding, the analysis in chapter 2 provides insight into the dimensions of adaptive capacity. In particular, three factors were identified that shape how adaptive capacity develops in a particular context. These are summarized, alongside the implications for adaptive capacity, in Table 3.1.

Table 3.1 The implications of resilience thinking for adaptive capacity

Insights from resilience thinking (chapter 2)	Implications for adaptive capacity
Opportunities for local livelihood changes are limited, constrained or regulated by processes that act at larger scales	Actors seeking local adaptation actions require influence over policies, processes and regulations at the district, national or even international scales
The perspective of actors at different scales are needed to build an understanding of complex systems and to respond to changes occurring at those scales	Adaptation decision making needs to integrate the knowledge of multiple stakeholders, including but extending beyond local actors
Links to cycles of experimenting and testing provide the learning necessary to respond to uncertainty and to prevent systems passing dangerous thresholds	The availability of appropriate local adaptation options will depend on access to information and knowledge gained from experimentation and testing

The overlapping implications of resilience thinking, summarized in the second column of Table 3.1, suggest that action on adaptive capacity breaks down into three interconnected areas:

First: the **power sharing** arrangements that are in place to expand communities' voice and influence over decision making (rows 1 and 2 in Table 3.1);

Second: the sources and processes that give rise to the **knowledge and information** that inform adaptation decisions (rows 2 and 3); and,

Third: the availability of **experimentation and testing** of adaptation options that are relevant at the local level (row 3).

These three dimensions provide a framework for structuring development support for adaptive capacity. As Figure 3.1 illustrates, the dimensions are linked and interdependent: the sources and processes that give rise to knowledge and information feed into power sharing relationships and emerge as collaborative actions – experiments and tests – that apply new understandings and produce learning.

One example of how the dimensions can be deepened is through the social learning processes explored in chapter 2. Social learning offers a model in

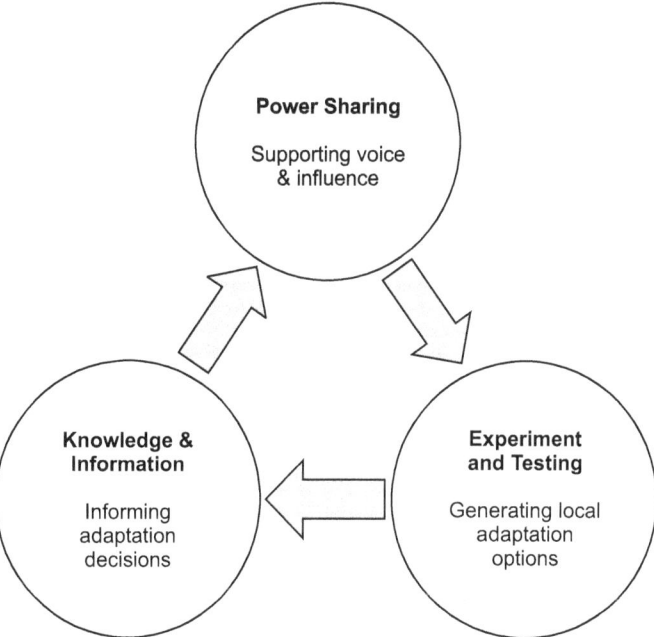

Figure 3.1 A framework identifying areas for action on adaptive capacity

which multiple participants from different scales are brought together (in power sharing relationships) to blend their different perspectives (sharing knowledge and information) and to undertake actions that generate shared learning (testing ideas in context, generating new knowledge and information). The result is a flexible, collaborative and adaptive governance system. But the particular manifestation of and – crucially – how to move towards social learning arrangements is highly context specific, depending on a range of factors including existing institutions, the mix and capacities of stakeholders, and the political context at local and higher levels.

The three dimensions are, therefore, focused on how to move forward with adaptive capacity in a particular development context. Attention is directed towards locally appropriate processes that can be supported and sustained to expand a community's power sharing, knowledge and information, and experimentation and testing opportunities – and thereby expand their capabilities in ways that enable them to engage with the challenges of complex systems – rather than the achievement of a particular governance arrangement. This approach recognizes that adaptive capacity cannot be achieved as a result of a single development intervention, but instead that a community's on-going processes of building and rebuilding relationships and networks can be supported in ways that help them to better meet the challenges of climate change.

The three dimensions – unpacked

The idea that there is an identifiable list of determinants that are sufficient to describe adaptive capacity is attractive, but illusory (Vincent, 2007). This is not least because adaptive capacity is context and scale specific, differing in nature for households, communities and nations. Measuring adaptive capacity is therefore beset with uncertainties, dependent not only on which indicators are identified as significant in a given context, but also on the nature of the challenge that is being faced. Indeed, as an attribute of a complex system, adaptive capacity is itself the product of the relationships and feedbacks that exist between people, policies, institutions and the environment in a particular place and time, making it variable and inherently unpredictable. Recognizing this, the intention of unpacking the three dimensions is not to enable the measurement of adaptive capacity or to facilitate a comparative analysis of different communities' ability to adapt. Rather, it is to identify the critical areas for action if adaptive capacity is to be addressed in a particular context. The purpose is to distil out key themes that development actors need to focus on if the capacity to respond to emerging and unexpected change is to be fostered. However, each of these dimensions overlap, require further unpacking, and are firmly situated within the existing local context. The following sections are intended to support a qualitative, contextualized analysis and assessment that leads to actions, but without seeking or suggesting a predetermined score sheet for adaptive capacity.

Power sharing

Power is significant because of the central role that it assumes in defining the opportunities and resources that communities can access. Recognizing power relationships is therefore an essential step in addressing adaptive capacity. Routinely, the control and use of environmental resources has been secured through the exertion of power to 'overcome, distort or impose upon more legitimate claims' (Borrini-Feyerabend et al., 2007: 51). Legitimate claims, such as those arising from a community's customary use or direct dependence on natural resources for subsistence, frequently carry little weight when pitted against well connected or better organized actors. Similarly, it is those who are best able to mobilize for collective action or take advantage of decision-making institutions that succeed in securing their interests in legitimate decision-making processes.

Competing claims and differing understandings of value are subject to unequal power and representation in social processes, undermining and excluding the poor while cementing their vulnerability – including to climate impacts (Ribot, 2009). The result is that the often quoted truth – that it is poor and marginalized communities that are most threatened by climate change – is true because relations of power, politics and economics have made it so. And within this context, women are frequently particularly vulnerable. Differences between women and men are often identifiable in terms of gendered livelihood roles, which in turn bring different risks from climate related shocks and stresses. However, women are also disproportionately vulnerable when they receive less education than men, or are excluded from political and household decision-making processes that affect their lives, or have fewer assets and depend more on natural resources for their livelihoods (UNDP, 2010a; Röhr, 2007; Box 3.2). Unequal gendered power relations can mean that adaptation strategies are pursued because, 'they reflect and reinforce gender inequalities, rather than because they represent the best adaptation choices' (Terry, 2009: 171).

The significance for those working to support adaptive capacity lies in recognizing that business as usual, in which decision making is exercised through existing institutions with inherent power relationships, will likely repeat and reinforce this pattern. As Nelson et al. observe, 'most adaptation does not necessarily reduce the vulnerability of those most at risk' (2007: 411). This is particularly the case as adaptations are not politically neutral: winners and losers are created as benefit is redistributed in a changed social-ecological environment, the more so when the result is transformations to established livelihood patterns. Political alliances and power relationships are therefore

---

Box 3.2 Gender and adaptation decision making

In the same way that gendered roles lead to differences in vulnerability between men and women, they also create opportunities for adaptation. Women are not just victims of adverse climate effects due to their vulnerability; they are also key active agents of adaptation. This is due to their often deep understanding of their immediate environment, their experience in managing natural resources (water, forests, biodiversity and soil), and their involvement in climate sensitive work such as farming, forestry and fisheries.

Women not only have roles as caregivers and nurturers, but also typically form strong social networks within their communities, thereby meeting a prerequisite for collective management of the risks posed by climate change. However, while their lives are typically closely tied up with natural resources, women are usually excluded from decision-making processes and thus barred from contributing their unique expertise and knowledge to the struggle to adapt to climate change.

The danger, of course, is that, if there is no gendered approach toward adaptation, these differences between men and women may be overlooked, inadvertently reinforcing gender inequality and women's vulnerability to climate change relative to men. Men's opportunities to promote adaptation originates from their limited childcare responsibilities; additionally, they are likely to have more education and greater professional and technical abilities, acquired from their working lives, that help them to adapt to climate change. The collection of sex-disaggregated data about such issues is essential to highlight the differences between men and women and to ensure that adaptation options are gender-sensitive.

If gender is overlooked in the planning of an adaptation intervention and women are not consulted, the measures may not be appropriate or sustainable. For example, women are often in charge of water management but, if they are not consulted about where to build new wells, the wells may be placed too far from the village, thereby actually increasing women's burdens.

Source: UNDP, 2010a: 20

inevitable in adaptation decision-making, played out not only in what gets decided, but who gets to decide (O'Brien et al., 2009). In this way, power directs us to question equity, both in process and in outcome: 'who decides what should be made resilient to what, for whom resilience is managed, for what purpose?' (Nelson et al., 2007: 410).

Moser suggests that development actors need to pay greater attention to understanding the 'social dynamics that underpin (motivate, facilitate, constrain) on-the-ground adaptation strategies and actions' in decision-making institutions, and specifically to address the 'value judgements and power dynamics embedded in adaptation decisions' (2009: 328). Jennings similarly draws attention to 'historically embedded and implicit power relations' and in particular notes the failure of indigenous or local environmental knowledge to penetrate bureaucratic thinking (2009: 245, and explored further in the next section). Ultimately, any local or NGO led action, however positive, can be undone by a policy environment that is outright hostile to or simply lacks a focus on marginalized or poor communities. Moser summarizes her discussion of decision making as a call to 'stop hand waving about adaptive capacity ... and increase our understanding of, and our ability to use or create more effective governance structures to realise it' (2009: 329). Networks of institutions, decision-making bodies and governance structures must, then, be subject to analysis – who represents whom, with what knowledge, with what motives – if the translation of adaptive capacity into adaptation is to benefit the vulnerable.

Analysing and understanding how power dynamics play out in different contexts therefore needs to be recognized as a task in its own right. Yet recognizing power means having an understanding of its characteristics. Borrini-Feyerabend et al. help here, identifying nine forms of power that shape the ability of individuals or groups to influence how natural resources are managed and utilized, and illustrating the multiple ways in which influence can be achieved (2007: 51):

- power of position (having authority, being in a position to make or influence decisions);
- power of knowledge (having information unavailable to others);
- personal power (being personally forceful, persuasive);
- household power (being from a well-connected family);
- group power (being a member of an ethnic, religious or other type of group that has a dominant social position or, for example, being male in male-dominated society);
- economic power (commanding financial and other economic resources in overwhelming amounts with respect to the resources of others);
- political power (having a powerful supportive constituency or access to political leadership);
- legal power (having strong expert legal counsel, or privileged access to courts);

- coercive physical power (having police or military backing or weaponry).

Peterson (2000) takes an alternative approach, directing attention to the different scales at which power can be exerted. Three categories emerge. First, 'overt' power influences decisions directly through force, incentives, or intimidation, operating locally and rapidly – in the 'here and now'. Second, 'covert' power is manifested in the manipulation of institutions, controlling whether issues are discussed or addressed. Covert power therefore operates at slower and larger, institutional, scales. Finally, 'structural' power is the slowest and broadest category, operating at the level of culture, restricting the set of issues or concepts that people think that they can make decisions over. Exposure to outside ideas and influences expands the local context of choice and can thus break down the control of structural power (Ensor and Berger, 2009b).

The value of Peterson's approach lies in bringing the perspective of resilience thinking to power, drawing attention to the less obvious, slower scales and the fundamental impact they can have on local decision making (chapter 2). Policy measures that set the rules and norms that govern institutions or control the freedom of communities to join or form new alliances, for example, are likely to be slow to change and may require development actors to focus on the non-local dimensions of power. Each scale needs to be unpacked if the local dynamics of decision making are to be understood. However, it is also important to guard against too rigid a categorization and recognize that the most influential actors may have the ability to work across scales. In a study of adaptive capacity in Nepal, Yates notes that 'powerful players and institutions are able to construct the adaptation needs of communities' by securing influence at multiple scales through horizontal connections to different government and non-governmental organizations and institutions (2011). Local actors may be able to employ, at different scales, some or all of the nine forms of power and so simultaneously exert control over how adaptation is defined locally and the perceptions of local needs at broader institutional scales.

The distinctive perspective of social network analysis helps in disentangling power relations across scales. By focusing on the connections between different actors, it reveals actors to be interdependent rather than autonomous, with relational ties (linkages) between them that are channels for the transfer or flow of resources. In this way, the network structure can be seen as providing the opportunities for or constraints on individual action (Wasserman and Faust, 1994). Network analysis seeks to systematically identify the nature and extent of the interconnections between actors within a network. It can be visualized as a web of connections that link diverse individuals and institutions, either directly or via other actors. For example, a household may be connected to other families in a village, a producers' association, a school, and a political organization. In turn, each of these actors has relationships with other parties that the household is able to indirectly access. The nature of these relationships will determine, for example, the household's knowledge of adaptation options, or its ability to exert influence on others. In this way, the analysis of social networks reveals a complex structure that governs the flow of material and non-material resources: the connections describe the access that different actors have to each other, whilst each link draws attention to the quality of relationships, revealing power in terms of social, political, cultural and economic resonances and barriers, and the interests or motives of the different actors.

Borrini-Feyerabend et al. summarize this by identifying that typically a 'complex patchwork of claims ... interplay with important power differentials within a context of relatively limited opportunities and resources' (2007: 55). Yet having recognized this complexity, they echo Nelson et al. (2007) in suggesting that the question in all cases boils down to one issue: how could the system be rendered the fairest possible? Their analysis, drawn from many different experiences, suggests that 'striving for equity' involves a sequence of three central issues:

- helping the underprivileged to develop their own entitlements, where entitlements are understood to be socially recognized claims to utilize, manage or take decisions over resources;
- recognizing entitlements rooted in valid and legitimate grounds (as defined by the relevant society) rather than entitlements rooted in the exercise of one or the other form of power;
- promoting a fair negotiation of functions, benefits and responsibilities among entitled social actors.

This perspective identifies an approach to tackling power relationships through development actions. In particular, it highlights the need to prepare the ground for community engagement in decision making before moving to focus on the institutions that deliver 'fair negotiation'. Developing and securing recognition for a community's entitlements (legitimate claims/ rights) is a necessary first step that demands opportunities and freedoms, including access to information, the ability to express views, time and resources to organize for action, and access to institutions without discrimination. Table 3.2 illustrates how these equity issues relate to a process of building communities' ability to engage in the co-management of natural resources. For the opportunity to gain entitlements to be built, attention is first needed locally so that legitimate claims and the barriers to their recognition can be identified in a particular social and political context (steps 1 and 2). Only then can attention turn to the appropriate institutional arrangements that enable negotiation and, in the best of cases, co-management between actors that share benefits and produce knowledge (steps 3 and 4). The need to understand power cuts across these issues, providing the foundation for strategies to secure equitable decision making and reworked institutions that are flexible enough to incorporate learning and are effective for the poorest.

Table 3.2 Steps towards power sharing

	Process	Equity considerations
Step 1	**Recognition** of the values, opportunities and risks associated with land and natural resources; **self-organization** to express those as own interests and concerns	Relevant **information** accessible to all; **freedom** of expressing views and **organizing** for action; time and resources to organize; fair system of representation
Step 2	**Recognition/negotiation** by society of the interests and concerns of the institutional actors as **'entitlements'** (customary and legal rights included)	Absence of social discrimination; **fair hearing** available to all institutional actors; political openness towards participatory democracy
Step 3	Entitled actors **negotiate agreements** and set-up **organizations**, **rules** and systems to enforce the rules to share natural resource benefits according to their own entitlements and capabilities	Existence of **negotiation platforms**; **capability** of entitled actors – including economic and political capability – **to negotiate** with others; non-discriminatory time, place, language and format of meetings; impartial and effective facilitation in languages all actors understand
Step 4	**Co-management partnership**: the relevant social actors share benefits and responsibilities amongst themselves; contribute knowledge, skills and/or financial resources; are held accountable for their agreed responsibilities; 'learn by doing' in natural resource management tasks	Acceptance of a measure of **democratic experimentation** ('legal and political space' to accept new actors, new rules and new systems to enforce rules); **flexibility** to adjust plans on the basis of experience; **effective enforcing** of negotiated agreements and rules

Source: adapted from Borrini-Feyerabend et al., 2007: 63

Yet while inequalities in power are unavoidable, power *sharing* necessitates moving beyond the identification of power as struggle, or the site of conflict between competing claims. The risk with a focus on power is that it is perceived as immutable and owned by different actors, rather than as determined through relationships that can be reshaped through collective processes (Collins and Ison, 2009: 362). Power relations undoubtedly limit the capabilities or 'opportunity set' of marginalized communities. But when capabilities are expanded through the sharing rather than wresting of power, complex systems can be tackled and benefit secured for and by poor communities. As chapter 4 discusses in more detail, there are several different approaches that, if embedded into development practice, can support the process of analysing and building relationships between actors and institutions in ways that open spaces for power sharing. Table 3.2 provides a schematic for these processes, suggesting a sequence of events in which actors and institutions that are increasingly remote from the local community are gradually drawn into negotiations and become increasingly prepared for power sharing. Throughout, facilitation is key, bringing together relevant stakeholders in reformed institutions that provide the opportunities (and release resources) for shared learning and new practices. It is these latter components – the emergent knowledge and activities – that are ultimately sought, delivering adaptation measures that are capable of responding to the complex, multi-scale realities of climate change.

Facilitating power sharing may require the intervention of 'bridging organizations' that are able to link local communities with actors at other scales – a role well suited to non-governmental organizations – as well as leaders that are able to shape change in organizations and build trust between diverse actors (Folke et al., 2005). Women may require particular and differentiated support if they are to contribute their perspective and distinct knowledge of natural resources, while identifying relevant stakeholders outside of the local community requires a comprehensive assessment of the local context. For example, in Mozambique, Osbahr et al. (2008) describe building relationships between those who influence the economic, social and political context of local activities. This encompassed the variety of actors whose interests overlap with the local community when agricultural adjustments are made within a season, or when drought requires the sale of livestock or taking of short-term loans, or when structural changes are necessary in management such as changes in land use, livelihood activities or crop type.

There are obvious challenges to power-sharing approaches, not least in securing the active participation of stakeholders, necessitating both personal commitment and, for those representing formal agencies, institutional support. It requires a recognition amongst those who are used to being decision makers that there is a need for a more inclusive process, and a readiness on their part to shift position, be influenced, and reach genuine consensus. Existing values and worldviews can be a barrier to progress: 'a legacy of past modes of operating combined with the persistence of outdated paradigms' mean that addressing complexity will often be a slow and uneven process (Adger, Lorenzoni et al., 2009: 6). The investment and skills required for facilitation, through which competing values and interests may have to be resolved, will often be considerable, while power imbalances between poor communities and their governing elites cannot be ignored. Stakeholders need to move beyond seeing institutions as 'frameworks for self-interest', to see them instead as guiding them in identifying their interests and shaping their environment (Goldstein, 2009). As Folke et al. point out, it is insufficient to rely on 'multistakeholder bodies that are often used by government agencies to increase legitimacy and manage conflicts without devolution of power' (2005: 450). Rather, it is through shared rights and responsibilities that resilience is promoted (Nelson et al., 2007).

The choice for adaptation is made clear by Few et al.: 'rather than retreating in the face of participatory difficulties, climate change adaptation needs to forge an honest and creative deliberative approach that both can be more democratic and can yield genuine benefits for the process of societal adaptation' (2007: 55). The challenge for those working to support adaptive capacity is the transformation of development practice towards these new forms of participation.

By seeking development actions that fulfil needs through power sharing processes, interventions can build relationships that lay the foundations for effective and equitable governance while securing immediate term needs. The challenges of creating and – ultimately – formalizing the space for reformed decision making will vary with context. But meeting these challenges is critical to moving beyond the historical power relationships that limit the options for change and to achieve a fuller reframing of the challenges of adaptation in the minds of all stakeholders (Collins and Ison, 2009).

Knowledge and information

Communities will need to expand their knowledge and access to information if they are to understand the challenges of an uncertain future and develop responses to the emerging impacts of climate (among other) changes. Bringing together different actors with a view to bridging their understandings of the world is therefore at the heart of adaptive capacity. In this way, the networks, institutions and decision-making spaces available to communities play multiple roles in adaptive capacity, helping to build a more complete understanding of the problem being faced, of the available solutions, and of the potential consequences and trade-offs. Yet unless climate scientists, bureaucrats and community members can come together, recognizing that each interprets and prioritizes the impacts of climate change differently, and collectively defining what successful adaptations may look like, then adaptation will remain partial and maladaptation likely.

The difference between knowledge and information is central to this view of adaptive capacity. Information can be understood to be organized data, used for communicating and reporting on the world. Knowledge, on the other hand, implies learning and the ability to use information. For example, where a weather forecast communicates information, knowledge would imply having previously used forecasts for a practical purpose and learnt from that experience. In this sense, knowledge makes information useful. But what is often overlooked is that it also means that we construct our knowledge on the basis of our (limited) experiences. One person's knowledge of a given issue will not necessarily be the same of another's. Power-sharing approaches can overcome this limitation by facilitating knowledge sharing and joint learning experiences. Through working together to better understand their situation, new, shared, ways of knowing are generated.

Knowledge and information can therefore be seen as an input to power sharing, as each actor brings their own experiences and multiple sources of information. But knowledge and information is also a product of the process, as new knowledge emerges and is made available (as information) to others. However, for the process to be successful, each actor needs to appreciate that their knowledge is partial and constructed through the prism of their experiences and worldview. In this way, power sharing processes provide participants with a window onto complexity, as each actor recognizes their knowledge to be incomplete and is enabled to form and revise their knowledge in the light of multiple perspectives and shared experiences.

Power, however, remains pertinent. The production of knowledge is part of the wider power dynamics that define the relationships between stakeholders, affecting who participates, who speaks and who benefits. Indeed, attempts at securing community participation in development processes have been beset with problems, critiqued for reproducing the views of the powerful rather than integrating local or indigenous knowledge (Jennings, 2009). This is particularly the case for groups that are usually excluded or marginalized from decision making, as it is their perspective that is least understood and least sought in conventional institutional processes. Yet 'social memory' and traditional knowledge are key components to be drawn out for adaptation, ensuring that the accumulated experiences held at the local level (in the form of customs, values or practice) feed into decision making, reducing vulnerability to surprises and connecting local knowledge and memory with visions of the future (Folke et al., 2005; Tschakert, 2007). Borrini-Feyerabend et al. summarize that (2007: 420):

> deliberative process, and the political negotiation over what constitutes valid knowledge in a particular context, deeply challenges professionals to assume different roles and responsibilities. In particular, citizens with professional knowledge will often need to shift to new roles that facilitate local people's analysis, deliberations and production of knowledge.

Climate information is one area where there is a well-identified need for professionals to engage constructively with local people (Harrison et al., 2007). Climate change is an emerging phenomenon with the potential to transform environments and challenge traditional expectations of seasonal patterns and climate extremes. An understanding of climate change predictions, different forecasts, and the associated uncertainty is therefore an essential component of adaptive capacity: as the Swedish Commission on Climate Change and Development point out, 'the uncertainties of climate change make ongoing access to information crucial' (2009: 19, see also Box 3.3). Expanding communities' networks through new participatory spaces can improve local access to climate information and, ideally, secure changes to climate information so that it becomes more relevant to the lives and livelihood strategies of the community.

New institutional processes can bring shared learning to the top-down, scientific terrain of climate knowledge. For example, Patt (2005) reports how workshops that built relationships between farmers and seasonal forecasters in Zimbabwe enabled the farmers to engage in forecast development. Those who attended the workshops were also found to be significantly more likely to adapt farming methods in response to the forecast information that resulted. Cash (2006) similarly recommends the use of 'boundary organizations' that work as an intermediary between local and scientific actors, whilst elsewhere Patt (2008a) describes the forming of 'partnerships' between scientists and users. In each case an essential feature is the investment in social capital to build bridges between the knowledge systems and priorities of different communities. However, a particular challenge of climate information lies in communicating the complexity that is inherent in climate modelling. The illusion of certainty and simplicity is attractive, raising the prospect of uncertainty being overlooked or underappreciated if information is exchanged without the investment of time and effort in building a shared understanding of climate change. An appreciation of complexity and uncertainty is essential if maladaptations or futile efforts at vulnerability reduction are to be avoided, meaning that equal emphasis is needed on developing a community's understanding of the science and scientists' understanding of a community's knowledge and needs.

Box 3.3 The role of forecasting information

Short-term forecasting and climate predictions both have a role in adaptation. Local knowledge, which traditionally underpins community responses to changes in the weather and seasons, needs increasing support in the face of unprecedented climate variability. The different challenges that climate change presents are best addressed through different approaches to forecasting or prediction. Extremely fast onset events, such as flash floods or cyclones, require weather forecasting (looking at the days or weeks ahead); fast onset events, such as drought, need seasonal forecasting (at a timescale of weeks and months); and slow onset events, such as changes to seasons or sea level, can to an extent be predicted through climate modelling (looking at multiple years into the future).

Yet there are also important links across these timescales, as changes to the climate place short-term events in a new context. Forecasting underpins the early warning that is necessary to ride out extreme weather events, be it through temporary relocation, securing property or ensuring adequate food supply. Climate change model predictions, on the other hand, may indicate the future likelihood of, for example, high rainfall or temperatures. This allows current unexpected events to be understood as part of an emerging trend (rather than as an anomaly) and offers a new context for infrastructure investment or livelihood decision making. This holds equally for incremental changes (such as gradual changes in temperature, sea level, or rainfall), which can go unnoticed, be perceived as anomalous, or be disguised in or misdiagnosed as part of cyclical climate variations. Such changes may be foreseeable through seasonal forecasting but must also be recognized as part of a new long-term trend if they are to be understood and planned for.

It is, therefore, through short-term and seasonal forecasting that adaptation actions are able to engage with the variable conditions that are superimposed on emerging climate trends. However, a risk remains that simple climate change messages can disguise the complexity that is inherent in climate modelling. This applies to uncertainty (where similar outcomes can be equally asserted) and the statistical nature of predictions (in that climate only offers a view of the average conditions over a long period). If the uncertainty in climate messages is overlooked or underplayed then actions may be driven towards maladaptations focused on predicted impacts that fail to materialize. Equally, however, complexity and uncertainty must not divert adaptation from addressing threats to livelihoods if such impacts are known and imminent.

In short, relevant and appropriate dissemination of climate change related information, encompassing weather forecasts, seasonal forecasts and climate change predictions, is an important – if not defining – feature of climate change adaptation.

Source: Ensor and Berger, 2009a

But climate is not the only area of new knowledge and information that is necessary for adaptation. If communities are to gain the capacity to adapt, there is a pressing need for access to information on how similar challenges have been met elsewhere. As drought emerges as a new issue in one location, for example, it is likely that communities in another have prior experience of livelihood strategies that would be viable. Alternative technologies, resource management approaches, crops or livestock are all likely to be available if appropriate links can be made between local communities and external knowledge holders. However, the take up and use of new information is strongly dependent on how that information is shared. This is in part because individuals respond differently to information and to knowledge gained through experience. Experience generally dominates in decision making, while even the most authoritative information generally has a role restricted to moderating the experiential response (Balstad et al., 2009). As a consequence, when new information is encountered through power sharing processes, in which experiences are shared, it can be effective in influencing changes to behaviour. Ensor and Berger (2009a), for example, report how the practice of vegetable growing was introduced to demonstration sites in Nepal in the first year of a project and, following successful cropping, spread within two years to almost all farmers in the area. Similarly, Gine et al.'s (2007) study of insurance take up amongst rural households in Andhra Pradesh reveals membership of social networks to be important in determining whether a new insurance scheme is adopted, as these networks provide opportunities for sharing information and advice. In both cases, the sharing of knowledge helps to reduce the real and perceived risk of adopting changes to livelihoods by observing and understanding the experiences of others.

A focus on knowledge and information means that power sharing processes are not simply the dry 'co-production' of knowledge through institutionalized encounters: they also encompass and demand visits, demonstration projects and opportunities to experience new technologies and livelihood strategies in ways that stimulate shared learning and localizes external knowledge. The richness of information sources and the ability to have a real, physical experience of otherwise abstract ideas will have an important role to play in securing informed adaptation choices.

Experimentation and testing

The testing of new technologies and methodologies in the local context links experimentation directly to the processes of knowledge generation described above. Stimulated by an awareness of climate change, new information is turned into new knowledge through its local application, with the learning that emerges providing the basis for a subsequent cycle of experimentation, testing and review. This is a process that is at the heart of adaptive capacity, rendering changes that are effective in context. Indeed, it has been suggested that the most adaptive societies are those with actors who have the capacity to experiment, and institutions in place to support them (Patt, 2008a).

The theory and application of participatory technology development (PTD) as an approach to embedding experimentation in the local context suggests that the ability to experiment can be understood in terms of *technical capacity* and *technology choice* (Murwira et al., 2000). Technical capacity refers to the skills and knowledge that individuals possess to engage in experimentation and technology development. Technology choice encompasses the ability to make informed decisions over alternative technologies. It results from the amount and quality of information that is available, and the emergent knowledge gained through shared learning and earlier choices. Skills and knowledge are therefore central to both technical capacity and technology choice, and can be mutually supportive: testing and experimenting with an improved range of technologies (greater choices) can yield new knowledge, skills and experiences (greater capacity) that can be applied again in the future.

This understanding of experimentation and testing adds to the demands on power sharing processes. Addressing choice places an emphasis on extending information sources through the breadth of stakeholders and the facilitation of visits, exchanges or trials. The characteristics of technical capacity draw attention to the inequalities in educational background, prior experience, support or assets that may exist between stakeholders – and that will need to be mediated – if local capabilities are to be extended. However, different aspects of power sharing processes can be brought to bear on the components of technical capacity. Through successive iterations, relationships that share power among stakeholders can be built, while developing the personal qualities and experience necessary for experimentation. Shared decision making that draws in formal institutions can also extend local access to and control over a diversity of resources, and flexibility in how and when they can be deployed to allow for experimentation. Yet significant challenges remain even if power relationships can be broken down and an environment for shared learning emerges. Attitudes and beliefs create a complex environment in which it is difficult to judge the readiness for risk taking that is necessary in experimentation. For example, evidence suggests that there is no correlation between farmers' wealth and their willingness to adopt new management practices (Phillips, 2003; Patt, 2008b) and that, while some social norms can stand in the way of experimentation, cultural attitudes may equally encourage experimentation. Marginalized communities have exhibited both conservatism and experimentation as a strategy to deal with environmental change (Ensor and Berger, 2009b; Patt, 2008a). Local, context specific facilitation is therefore necessary if appropriate support for experimentation and the local application of new ideas is to be provided.

Examples from PTD practice show how this can be achieved in different settings, in particular by working with progressive community members and rapidly demonstrating the value of experimenting (chapter 4; see also Murwira et al., 2000; Lewins et al., 2007; Ensor and Berger, 2009a). However, support for experimentation can come in many forms. Local extension services and the availability of technical training provides a basis for the local adoption and adaptation of new ideas, while initiatives such as buyer groups, cooperatives or seed saving schemes can provide access to a greater degree of security or risk sharing and thereby the freedom to experiment. But through building new networks of relationships for technology choice and technical capacity, PTD activities go a further step in supporting adaptive capacity, by providing a basis from which subsequent spaces for power sharing can emerge. Moreover, by working to improve the capacity of communities to make informed decisions over their own technology choices, the methodology of PTD specifically operates to minimize the risks of inappropriate technology development and build the capacity to make future adaptations.

Supporting adaptive capacity

Development support for the three dimensions outlined above will not directly deliver adaptive capacity. The dimensions do, however, provide a framework of processes that can be supported through development actions, so that individuals and communities are able to rework their relationships in ways that generate shared learning through shared power. By focusing on power sharing, knowledge and information, and experimentation and testing, development can support the capacities and opportunities that communities need to address the challenges of climate change as they emerge.

Specifically:

- Power sharing means forging new, equitable relationships between actors with different backgrounds and experiences. It emerges as a component of adaptive capacity from an acknowledgement that the impacts of and responses to climate change are played out in complex social-ecological systems. This means that the perspectives of different actors are necessary to build an understanding of how the system changes and reduces the chances of crossing thresholds. Yet to be successful, these processes must acknowledge and move beyond the power relationships that are frequently fundamental in sustaining impoverishment and closing down the opportunities for adaptation in marginalized communities. Supporting adaptive capacity thus necessitates a focus on the networks of relationships, rights and duties through which power is mediated in a particular context – but with a view to bringing actors together to share power in reworked institutional processes. Power sharing is therefore the first, and overarching, area of focus, bringing processes, institutions and new forms of participation into the foreground of development work.
- To be effective for adaptive capacity, work on building power sharing relationships must be combined with a focus on the emergence of shared knowledge and access to new sources of information.

Knowledge and information are central to adaptive capacity, as a full understanding of the challenges and opportunities faced is necessary to support adaptation decision making. However, communities will need to integrate new sources of information in order to address the emerging challenges of climate change, while combining knowledge from different worldviews to address complexity. Attention is therefore drawn not only to access to information, but also to how knowledge and understanding can be generated through relationships that support shared learning.

- Experimentation and testing interlocks with its complement, knowledge and information. Knowledge and information on adaptation strategies can be tested and experimented with locally, thereby producing new knowledge that can inform decision making through a contextualized understanding of alternative responses to the challenges of climate change. Securing processes of experimentation, testing and learning enhances resilience by preventing social-ecological systems from becoming too rigid or fixed in the face of uncertainty, but means that power sharing relationships must be sufficiently flexible to respond to emerging knowledge and able secure resources and support for local actions.

The overarching focus on power sharing does not immediately translate into a demand for new governance structures or institutional arrangements. Rather, it implies an investment in time to build confidence and to identify shared challenges, first within a community and then with more remote stakeholders. The emphasis is on transforming a community's perception of their needs into legitimate claims that can be defended through negotiation, and on skilful facilitation to enable those with competing interests to reach agreement. This is a view of participation that focuses in on developing meaningful relationships in fora that are at scales appropriate for inclusive, shared decision making. Knowledge and information ultimately yield adaptation options, but flow from these shared processes via the learning from experimenting and testing new knowledge. Chapter 4 explores how these different elements can be achieved in practice, focusing on empowering communities to secure opportunities for change through engagement and influence at the local and broader scales.

CHAPTER 4
Adapting development – in practice

This chapter moves on from the question of what constitutes support for adaptive capacity, to how that support can be provided in practice. Accordingly, the focus turns to lessons, tools and methods that can be called on from within development that relate to the different components of the framework presented in chapter 3. The intention here is not to suggest that there is definitive list of methodologies or strategies that, if adopted, enable development actors to work in support of adaptive capacity. Rather, the purpose is to demonstrate that there is a body of experience of working with power sharing, knowledge and information, and experimentation and testing that exists within the development sector.

As power cuts across the framework presented in chapter 3, the next section draws out four lessons derived from experiences of working directly with power issues. Rights-based approaches brought power into the foreground of development practice, providing insights into how the poor and marginalized can be supported to make claims as the first step towards power sharing. In the following section, participatory action plan development (PAPD), a consensus building approach, is presented and explored through a detailed case study. Consensus building is a common thread through several different methodologies that bring stakeholders together in power sharing relationships for joint planning, knowledge sharing and learning. Finally, an example of participatory technology development is used to show how experimentation and testing can be supported by building capacity, relationships and knowledge that are relevant for the local context.

This chapter concludes by emphasizing that supporting adaptive capacity means working on all elements of the framework in a coherent, programmatic manner. The overarching goal is not one of simply implementing the three dimensions of support for adaptive capacity, but of working with communities so that they can become self-organizing, learning and adaptive, and thereby able to sustain themselves in the pursuit of their values and priorities.

Rights-based approaches: lessons for adaptive capacity

Chapter 3 locates power at the heart of adaptive capacity: it cuts across the framework and is explicitly the focus of the overarching *power sharing* component. Equity (in terms of fairness, representativeness, and the incorporation of diverse values and views) – and effectiveness (in addressing complex systems, as described in chapters 2 and 3) – means that specific support is needed to ensure that the most vulnerable participate in decision making. Addressing their routine exclusion from these processes demands development attention to the determinants of marginalization, including culture, gender, ethnicity and, ultimately, power. Yet a call for this perspective is not new: since the emergence of development as a sphere of activity, it has been shadowed by a narrative that looked beyond economic indicators and called for development attention to be directed to human rights and social justice. For example, the International Commission of Jurists, summarizing in 1981 a body of worked that they had started in 1959, called for 'structural changes at all levels – local, national and international, that would enable those concerned to identify their own needs, mobilise their own resources, and shape their own future in their own terms' (ICJ, 1981). Rights-based approaches to development emerged from these ideas, bringing into the mainstream a movement that had long recognized the interconnectedness of impoverishment and disempowerment (Gready and Ensor, 2005a).

In this view, development is a process of transforming relationships through actions that amplify the voices of the poorest. Power holders are identified and seen as duty-bearers with the responsibility to fulfil rights, enabling the 'development of claims that seek to empower excluded groups and that seek to create socially guaranteed improvements in policy' (Uvin, 2004: 163). Through a renewed focus on the political dimensions of development, rights-based approaches have influenced the emergence of a body of process-orientated thought and practice that is relevant to addressing power, supporting the legitimate claims of the poor and working towards power sharing relationships, each of which are central to the first dimension of the framework developed in chapter 3.

This is not to suggest that rights-based approaches on their own are an answer to the challenge of supporting adaptive capacity; rather, it is that they have a useful role in framing an alternative perspective on development. In this, the rhetoric of rights – of politics, power and accountability – is important, as the structural problems that adaptive capacity must overcome are as much a challenge to the mindset of development actors as they are to practice. The issue is how best to frame development so that it turns its attention to the capacity of the poorest to engage with power: in this right-based approaches offer four lessons for those working to support adaptive capacity. These lessons build on Gready's analysis of the contribution of rights-based approaches to development. They relate to how communities can be supported in the process of developing power sharing relationships and, in particular, to working with communities to identify, express and claim their legitimate entitlements (2008).

Lesson 1: participation should be transforming

Where applied, rights-based thinking has given rise to a shift in perspective on what constitutes development. In adopting a rights-based approach, the principles embodied in human rights are integrated into and upheld through development actions. The impact has been to bring rights into the heart of development, informing and reconceptualizing principles and practice. Relating this perspective to climate change issues in development, the Office of the High Commissioner for Human Rights summarizes that (2009: 81):

> looking at climate change vulnerability and adaptive capacity in human rights terms highlights the importance of analysing power relationships, addressing underlying causes of inequality and discrimination, and gives particular attention to marginalized and vulnerable members of society. The human rights framework seeks to empower individuals and underlines the critical importance of effective participation of individuals and communities in decision-making processes affecting their lives.

These two strands – analysing power relationships and working towards empowered participation – are fundamental to rights-based approaches. They provide development with a rhetorical framing and programmatic focus that draws attention to the absolute necessity of engaging in processes that enable poor and vulnerable communities to define and demand their own development priorities. From this perspective, the pressing need is to understand and address the barriers to participation in decision making, so that progress can be made towards power sharing arrangements that enable those affected by climate change to define their own solutions.

Molyneux and Lazar, in examining the role of rights in development projects in Latin America, describe how this view of participation translates into a process of 'changing mentalities', in which those who express their needs move from a focus on charity and favours to being claimants with legitimate entitlements. Rights are used as a mechanism to encourage a mode of participation that at best will 'empower the poor to analyse their own personal situation, attribute responsibility and work out the means to improve it' (2003: 9-10). Awareness raising and accountability then become part of development practice, as the focus on processes shifts development interest away from the provision of services, and onto actions that enable communities to identify those responsible for their delivery. As a consequence, actions to secure changes in the social and political context that provide communities with their development opportunities become a central aspect of rights-based programming.

Those working to support adaptive capacity similarly need to focus on participation with the goal of transforming attitudes and understandings in ways that enable communities to identify and overcome the challenges they face in their local context. In a review of rights-based practice, Gready summarizes that this amounts to a reframing of participation towards 'a focus on advocacy and mobilisation that potentially nurtures inclusive problem-solving, citizenship and political activism ... often most concretely achieved via a linkage with agency and empowerment' (2008: 742). If communities are to move beyond existing power relations, they will need support that enables them to change their relationship with power holders so that they can advocate for and engage in decision-making spaces. Rights based approaches systematize this through programming that (Nelson and Dorsey, 2003: 2017):

- prioritises the most vulnerable and their unfilled rights in programme design;
- integrates education and awareness raising of rights and duties into projects, enabling communities to identify their legitimate claims;
- involves affected people in project design, implementation and oversight processes, by right; and
- supports accountability, by enabling communities to hold duty bearers responsible.

Lesson 2: adaptive capacity is a political issue

Rights-based approaches, as described above, emphasize causal analysis and social actions that amount to the politicization of development (Gready and Ensor, 2005a). A rights focus demonstrates that poverty is a consequence of structural failings in society – of rights denial. Development is not, then, about charity or technical assistance, but is a process of identifying and rectifying the underlying causes. Poverty is recast as impoverishment, identified as an act that is done to people: poor 'is not what people are, but what they have been made' (Mander, 2005: 240).

This mindset is equally relevant for adaptive capacity. If the components of adaptive capacity include processes that provide access to participation in decision making, then supporting adaptive capacity is about taking sides in issues that are political rather than technical in nature. The structural inequalities and power relationships that underpin exclusion from economic opportunities and political processes are brought to the foreground, and attention is focused on the cultural, ethnic, religious and gender dimensions of marginalization. As a consequence, working with either rights-based development or adaptive capacity means 'empowering marginalised groups, challenging oppression and exclusion, and changing power relations' and as such it means challenging established structures within society, falling 'squarely in the political realm' (Uvin, 2010: 172).

This perspective was recognized in a 2008 symposium convened to explore the role of resilience in development (Leach, 2008: 15):

> resilience thinking ... will often require politically progressive thinking and action to challenge and transform unsustainable structures and framings in radical ways, and to hold powerful actors and networks to account. Depending on the issue and the setting, strategies might involve a spectrum from discursive and deliberative politics, to more antagonistic politics of resistance and struggle; all involve moves away from the managerialism that characterised early resilience approaches, towards conceptualising it in fundamentally political terms.

The implication is that adaptation 'depends on human agency, including the role of individuals, collective movements, leaders, and institutions, and it often involves political struggle' (O'Brien et al., 2009). This, then, is the basis of work on adaptive capacity. Different forms of engagement with authorities, power holders and institutions at different levels to enable shared decision making lie at the heart of building the relationships and processes through which adaptation resources are secured. Taking sides in the way that rights-based approaches demand implies identifying power holders and duty bearers and paves the way for political activism. In some circumstances, entering political processes may be the only option to secure the policy changes necessary for adaptation, as in Niger where those working with the Tamasheq pastoralist community conclude that to secure land tenure reform, necessary to protect regenerated pasture, the community 'must take charge and enter the political process' (Ensor and Berger, 2009a: 130).

But the identification of rights holders and duty bearers does not necessarily lead to a confrontational approach. In many circumstances where rights are employed in development, attention is drawn towards modes of communication that are empowering, two-way and interactive processes, enabling those with claims to identify desired changes and negotiate with power holders (Jonsson, 2005). Here, rights realization is 'triggered by the process of communication; that is by an interaction between claim holders and duty bearers that admits the former into the decision making process' (Gready and Ensor, 2005a: 25). This understanding deliberately moves beyond seeing rights purely as a site of conflict with power holders, as is often assumed when rights are seen as a legal tool. Instead, a rights focus directs development work towards power sharing and the building of capacities and support for equity in decision making. In the same way, recognizing adaptive capacity's political dimension does not imply that confrontations with power are desirable. Collaborative methodologies are considered in more detail in the next section, but can be seen as approaches to re-imagining participation that, through building productive relationships with identified duty bearers, places those with claims at the centre of the process of securing outcomes.

Lesson 3: the state must be accountable

Accountability is a natural consequence of a focus on rights. Linking rights holders and duty bearers in social, political or legal processes provides different mechanisms for holding those with responsibility to account. Power and politics is never far in the background, and in some circumstances overt political activism may be necessary, such as through documenting and reporting on the impact of policy on marginalized communities. 'Speaking truth to power' in this way has emerged through rights-based activism and may be a valuable strategy in securing a voice for poor communities or enabling the monitoring of state-led adaptation actions. Yet, as noted above, the consequence of a rights focus is not limited to confrontational approaches, and effective action may be achieved through strategies that support and empower duty-bearers. Many reported examples of rights-based approaches adopt this approach (Gready and Ensor, 2005b), and capacity building with institutions and local governments to support engagement with poor communities and the delivery of services is a strategy adopted by many development actors. Working in this way to build accountability in institutions will be a key strategy in support of adaptive capacity, and is explored in the following section.

The attention to power holders and duty bearers in rights-based approaches also brings focus on the state as a key actor. Its role in development is highlighted and brought centre stage through its responsibility to respect, protect and fulfil rights, as is the need to develop strategies for engagement, be it through political activism, institutions, or challenges through the courts. Just as the modes of engagement with the state vary with context, so the state itself has multiple faces with which development actors and communities can negotiate – but participation in these different decision-making processes is a right, not a transitory gift.

The implication is that, for adaptive capacity, the state has ultimate responsibility for ensuring that adaptation support and outcomes are achievable at the local level. Accountability in its many forms directs attention onto the success or failure of the state in this role, and is a tool that civil society must grasp to help ensure that the state continues to meet its obligations to the poorest and most vulnerable.

Lesson 4: the law may be a tool for adaptive capacity

Rights-based approaches also draw attention to the legal avenues to hold responsible authorities to account when they fail to meet their obligations, as laid down in international, regional or national human rights provisions. The presence of legal norms has been used in rights-based development in two ways: through the direct use of the law to make claims in courts, and the 'strategic' use of the law (Gready, 2008: 739). The more widely rights are recognized, the greater the prospect for political, rather than just legal, action. While direct legal action can bring impacts, it can be a slow process that demands significant resources, often without a clear prospect of success. For widely accepted rights, however, knowledge of the law can be used strategically to effect political change through campaigns that raise awareness and build pressure on decision makers.

The direct application of the law refers to using legal methods to challenge abuses of power that can be framed as rights violations. In many cases this will involve in engaging with economic, social and cultural rights (ESCRs), such as in the right to food campaign in India (Right to Food Campaign, 2011). While this is a difficult and in many cases new area of law, there is an emerging jurisprudence that explores the implications of many economic and social rights. Moreover, there is evidence to suggest that there is a role for the law in securing state action on environmental issues. The implication for those working to support adaptive capacity is that there is scope to frame barriers to adaptation in terms of a failure of the state to meet its obligations to its citizens. Knowledge of rights law and access to legal recourse, where desirable, then become tools that enable communities to make changes through a rebalancing of power in favour of the most vulnerable.

The Office of the High Commissioner for Human Rights (OHCHR) directs attention to two areas of potential for legal action in relation to climate change. First, it is noted that the 'jurisprudence of regional human rights courts has underlined the importance of access to information in relation to environmental risks' (2009: 78). As the framework presented in chapter 3 makes clear, access to information is a central pillar of adaptive capacity and, as a widely recognized right, there is good potential for framing a demand for climate information in rights terms. Second, the OHCHR also notes that several claims of environmental harm have been considered in different bodies, from national courts up to the Human Rights Committee of the UN. These actions on rights, including to life and to health, have shown that 'the matter of the case would rest on whether the State through its acts or omissions had failed to protect an individual against a harm affecting the enjoyment of human rights.' (OHCHR, 2009: 73).

An important aspect of a rights framing is that, regardless of the additional burden that climate change places on public finances, 'States remain under an obligation to ensure the widest possible enjoyment of economic, social and cultural rights under any given circumstances. Importantly, States must, as a matter of priority, seek to satisfy core obligations and protect groups in society who are in a particularly vulnerable situation' (OHCHR, 2009: 77). This 'core obligation' provision is a key pillar in how economic, social and cultural rights (ESCRs) have been interpreted. The UN Committee on Economic, Cultural and Social Rights has authoritatively elaborated on the nature of state obligations, explaining that each state party to the convention must provide the 'minimum essential levels of each of the rights', such that it must be able to 'demonstrate that every effort has been made to use all resources that are at its disposition in an effort to satisfy, as a matter of priority, those minimum obligations' (OHCHR, 1990: 10). Furthermore, 'even in times of severe resources constraints ... the vulnerable members of society can and indeed must be protected by the adoption of relatively low-cost targeted programmes' (OHCHR, 1990: 12).

ESCRs therefore provide specific tools for those working to challenge entrenched power relations that undermine the adaptive capacity of the poor and vulnerable. Access to land and natural resources, for example, have been identified in many contexts as significant constraints on adaptation options (CAPRi, 2010). A rights focus casts this challenge in a new light, seeing these goals as rights unfilled and defining, through the work of international and regional courts, treaty bodies and expert opinion, minimum standards that must be achieved by states. There is not only an obligation on states to take measures towards their fulfilment ('progressive realization'), but also to avoid reduction in the degree of satisfaction of those rights ('non-regression').

Finally, however, a note of caution is necessary, as the work of the courts on ESCRs remains controversial and yields recommendations that have proved difficult to translate into policy actions (Gready, 2008: 738). It can also be naïve to assume that state organs such as the courts are blind to the prejudices that underpin poverty: 'the potential of the justice system to sustain and reinforce discrimination (and therefore poverty) [is often] overlooked' (Tomas, 2005: 172). Recourse to the direct use of the law therefore requires specialist legal advice and support, is potentially costly, does not guarantee success and may not be preferable when strategic actions are a realistic alternative.

Consensus building: sharing power and knowledge

While the lessons from rights-based approaches frame development in a way that helps to understand the challenge of supporting adaptive capacity, consensus building approaches provide a suite of tools that development actors can apply. 'Consensus' implies reaching a negotiated agreement in which all stakeholders are satisfied with the outcome. While this does not require total agreement, it does bring all stakeholders to the point where none have concerns or objections that they feel are significant enough to justify blocking the shared wishes of the group. As such, carefully facilitated processes between representatives of diverse interests or perspectives can reach consensus and share ownership of the decision. Power sharing arrangements such as co-management or collaborative agreements are defined in this way, relying on consensus rather than a majority vote and generating an outcome that is based on an agreed balance of interests between informed, entitled and engaged stakeholders.

Consensus building approaches are therefore an important practical avenue towards negotiated power sharing arrangements, as the following examples demonstrate. Recognizing imbalances in power, relationships are developed between diverse interest groups by sharing knowledge and developing a common understanding of the problems being faced. This is achieved through a simultaneous focus on four key elements: stakeholders and their relationships, the co-production of knowledge, the development of conducive institutions and policies, and – crucially – sensitive and timely facilitation (Collins and Ison, 2009).

Participatory Action Plan Development (PAPD)

Participatory Action Plan Development (PAPD) is a consensus building tool for use with marginalized communities that helps to build new political and institutional relationships. It emerged from research in Bangladesh on the livelihood challenges of those living in the floodplains and combines several tools that have emerged as good practice in participatory rural appraisals (PRAs). Practical Action modified the approach for use in the charlands of Bangladesh (erosion-prone accreted land that forms river banks or river islands), an experience that is reported in depth in the book *Voices from the Margins* (Lewins et al., 2007). A decade of working with PAPD in large natural resource management projects in Sudan, Bangladesh, India, Cambodia and Vietnam was recently summarized in the form of a facilitator's guide, published by Practical Action as *Consensus building with Participatory Action Plan Development* (Taha et al., 2010). These sources in particular support the overview of PAPD presented here.

PAPD is based on six key activities that provide a structured and repeatable approach to help people identify their shared problems and the potential pathways to their resolution. Acknowledging the fact that power relations shape local opportunities for community-based development, it seeks to develop the conditions for power sharing between the multiplicity of actors and interests that are present. As a consensus building approach, the emphasis is on building relationships between diverse stakeholders to raise awareness and understanding of their different perspectives, using skilful facilitation to ensure the full participation of the most vulnerable and to avoid descending into a conversation that follows the fault-lines of familiar local animosities. As Lewins et al. summarize, PAPD is 'a methodology to build local consensus by uncovering co-dependencies and developing greater understanding between stakeholders', centred on (2007: 21):

- a recognition of the range and connectedness of livelihood interests within communities;
- acknowledgement of the role of relationships, trust and institutional support in collective action;
- an understanding of group dynamics and the value of a well-facilitated and punctuated sequence of tasks and achievements; and
- the use of simple participatory tools with participants.

The six phases of the PAPD methodology are described in brief in Box 4.1. Attention is given to both primary and secondary stakeholders, with the primary stakeholders involved throughout. Primary stakeholders are those that rely directly on the natural resource base for their livelihoods, whereas secondary stakeholders are 'those people or institutions that have a less direct stake but also have an interest and role to play in management decisions affecting natural resources and livelihoods' including, for example, extension agents, technical bodies associated with agriculture, fisheries or water, or local and regional government bodies (Taha et al., 2010). Use of the two groups deconstructs the concept of 'community' by identifying actors with different but overlapping livelihood concerns and interests. A checklist of actors that may be relevant is provided in Box 4.2, illustrating the breadth of stakeholders that may need to be included in efforts to secure effective power sharing. Network analysis, discussed in chapter 3, helps to visualize the direct and indirect relationships and power differentials between stakeholders. 'Pattern analysis', often applied in rights-based approaches, can also be used as a tool for understanding the complicated interplay (or pattern) of rights and duties that exist in a particular context,

Box 4.1 The six steps of PAPD

1. Preparation

The *purpose* of this stage is to help facilitators develop sufficient background knowledge of local issues with primary stakeholders and to help them select a representative group with the community.

The *approach* is to seek the support of the community in developing a profile of the area and in participating in future planning.

The *process* includes community profiling and participant selection

2. Problem census and problem prioritization

The *purpose* of this stage is to initiate discussion of the range of livelihoods and natural resource management problems in the area, to increase awareness of these issues and their underlying causes, and of how they impact different groups. Discussing the concerns of each stakeholder group, in turn, helps build confidence and empathy between the primary stakeholders.

The *approach* is to seek input from each group of stakeholders in turn and to discuss the relative importance and root causes of specific problems across the whole group.

This *process* commonly involves separating the participants into smaller stakeholder groups (e.g. women, farmers, fishers) to work with one facilitator and list what they view as the major livelihoods problems. Each group then reports its findings to the whole group.

The facilitators must help the participants identify the most commonly mentioned problems and to group or cluster them into themes. Agreement must be reached to discuss three or so priority problems in detail. To keep the discussion focused, and to move towards potential solutions, a 'cause and effect analysis' can be conducted, documenting in table form the causes, impacts, affected groups and potential solutions for each problem.

The *outputs* are: 1) broad agreement to target one key problem that affects local stakeholders; 2) understanding of the root causes and potential solutions to this problem, and; 3) a written or photographic record of this planning.

3. Information gathering

The *purpose* of the information gathering stage is to increase local awareness and to form new links with secondary stakeholders. Before local stakeholders can make informed choices about potential solutions to problems they need to seek information and advice. This is the first time in PAPD that secondary stakeholders are engaged.

The *approach* is to support a small but representative committee to gather information about the problem or problems prioritized.

The *process* involves the committee consulting relevant stakeholders like NGOs, government, tribal leaders and others such as households highly involved with the issue. The committee will also make an initial assessment of the scale of the problem and reflect on a possible timescale for the PAPD process.

The *output* is increased awareness of the issue and written documentation of the advice and support received by the PAPD participants.

4. Analysis of solutions

The *purpose* of this stage is to work towards realistic solutions to the problems and to build awareness of the types of additional stakeholders that may need to be engaged.

The *approach* is to analyse in detail the feasibility of alternative solutions with the whole group.

This ***process*** relies on a tool called 'STEPS analysis'. This tool allows the facilitators and the group to consider all aspects of proposed solutions in relation to social, technical/financial, environmental, political/institutional and sustainability factors. For example, the social factors may include discussion of which groups might be affected positively or negatively by the proposal, and how the negatives may be minimized. This requires group discussion with broad input and debate.

The ***output*** is a STEPS table for each key problem tackled.

5. Public feedback

The ***purpose*** of this stage is to communicate the planning process and its achievements (key problems and carefully analysed solutions) to the wider local community and sets of relevant secondary stakeholders and influential people. The intention is to seek broad consensus and the technical, financial, political and institutional support needed to turn the proposal (or proposals) into actions.

The ***approach*** is to make public the findings of planning by explaining both the process undertaken and the draft plans. The facilitators should invite the opinion of the audience and should listen carefully to accommodate their comments.

The ***output*** will be to have achieved support and to have received the opinion of the PAPD plans from the wider community. The facilitators should document issues arising from the meeting.

6. Action plan development and implementation

The ***purpose*** of this stage is to provide support to the local communities to finalise a working plan and take the first steps towards implementing it.

The ***approach*** is to set up discussion and negotiation between PAPD participants and supportive stakeholders whereby a detailed work plan for the proposal can be made.

The ***process*** involves facilitating a meeting to review the draft plans and STEPS analysis with relevant technical service providers, PAPD participants and supporting committees (these may include existing project committees or regional networks). The facilitators will encourage the group to identify realistic schedules and stages to implementation, paying specific attention to technical, institutional, political and financial requirements. Agreement must be reached on specific roles and responsibilities and in particular the structure, membership and function of any group formed to coordinate implementation of the plan.

The ***output*** will be a detailed implementation plan.

Source: adapted from Taha et al., 2010

Table 4.1 Pattern analysis for a hypothetical fishing community

Duty bearers \ Claimants	Small-scale fishers	Local Authority	District Extension Office	National Government
Local Authority	Fair grants of fishing licenses		Access to all fishery sites	Report income from licenses
District Extension Office	Extension support for small scale fishers	Training for licensing staff		Comply with national fisheries policy
National Government	Change policy to allow small-scale fish species onto national market	Funding to enforce licenses	Salaries and costs for extension workers	

and thereby which actors are implicated if poor people's entitlements are to be recognized within equitable power sharing arrangements. By identifying where responsibility lies for providing services or opportunities, pattern analysis can help in identifying important secondary stakeholders, as illustrated in Table 4.1 for a hypothetical fishing community.

Three features of PAPD are of particular significance for supporting adaptive capacity. First, the focus of problem analysis and prioritization (step 2, Box 4.1) is on drawing out common elements in the concerns and interests of the different primary stakeholders. This builds a shared sense of purpose between community members that may previously have identified themselves in terms of competing interests, paving the way for power differentials to be overcome through the uncovering of mutually supportive relationships. It can also build the confidence of marginalized groups, such as women, to express their interests by providing the opportunity to meet and discuss issues together, and opening a space in which they have a platform to communicate their priorities to other sections of the community. An emergence of shared interests and a common voice that includes the poorest provides the grounding necessary for power sharing arrangements that enable the development and implementation of action plans. Here, the opening steps of PAPD reflects those identified in Table 3.2 in that they lay the foundation for co-management and benefit sharing.

The second important feature is that links between primary and secondary stakeholders are forged through the process, in particular in the information gathering phase (step 3) in which the emphasis is on building relationships to secondary stakeholders. As chapter 3 proposes, it is essential for supporting adaptive capacity that communities expand their networks to secure the support of actors at higher scales and for decision making to integrate the perspective of multiple stakeholders. Finally, through the six steps, PAPD facilitates knowledge sharing on the focus issue(s) between a wide range of actors: representative primary stakeholders from different interest groups, secondary stakeholders, the wider community and influential individuals.

Taken together, these features of PAPD combine to build horizontal and vertical relationships in an environment that is designed to highlight the value of collaboration. Through the process of building local consensus around an issue of significance to the community, new institutional and social relationships are formed that open up opportunities for the poor and provide 'a foothold for longer-term resource use negotiation, committee formation and community-based management' (Lewins et al., 2007: 23). Support for adaptive capacity emerges through these features of PAPD because, rather than focusing on the particular physical end point in the project (a building, for example), working through PAPD demands that development actors engage with the process itself, learning to recognize and navigate obstructions to progress towards a consensus-based plan.

> **Box 4.2 Possible stakeholders in natural resource management**
>
> - Local actors, including the communities, organizations, groups and individuals who live and work close to the resources, the ones who possess knowledge, capacities and aspirations that are relevant for their management, and the ones who recognize in the area a unique cultural, religious or recreational value. (This is an ample category, including several sub-categories.)
> - Natural resource users, including local and non-local, direct and indirect, organized and non-organized, actual and potential users, as well as users for subsistence and income purposes.
> - National authorities and agencies with explicit mandate over the territory or resource sectors (e.g. ministries or departments of forests, freshwater, fisheries, hunting, tourism, agriculture, protected areas and, in some cases, the military).
> - Sub-national administrative authorities (e.g. district or municipal councils) dealing with natural resources as part of their broader governance and development mandate.
> - Non-governmental organizations and research institutions (e.g. local, national or international bodies devoted to environment and/or development objectives) which find the relevant territories and resources at the heart of their professional concerns.
> - Business and industries local, national or international (e.g. tourism operators, water users, international corporations), which may significantly benefit from natural resources in the area.
> - Non-local actors, national and international, indirectly affected by local environmental management (e.g. absentee landlords, down-stream water users, environmental advocates or animal rights groups).
> - Individual professionals employed in environment and development projects and agencies dealing with the management of natural resources in the area.
>
> *Source*: Borrini-Feyerabend et al., 2007: 41

Case study: PAPD in the Bangladesh charlands

Practical Action's experience with PAPD in Bangladesh illustrates many of these points. The work took place in two villages in the charlands of Bangladesh, where the majority of communities are recently displaced due to the regular shifting of the riverbank and island land. In this environment, social capital is low, contested land rights underpin local conflicts, communication and access to markets is poor and government services are very weak or non-existent. Chronic poverty is the norm. The villagers' concerns centred on supporting their livelihoods and food security, and they identified secure access and fishing rights at the local water body as a priority issue from the start.

PAPD's facilitated approach was applied by Practical Action to promote collective planning and management as a way to resolve conflicts and support the interests of the poorest around the fisheries. While the six steps (Box 4.1) were applied flexibly, they provided a project design structure that satisfied the four conditions for the emergence of co-management discussed in chapter 3 (recognition of different values and self-organization to express interests; negotiation and recognition of legitimate entitlements; negotiation of agreements, organizations and rules; and, implementation of a co-management partnership (Table 3.2)):

1. The starting point for the PAPD process was the recognition of different values attached to the water body. The problems with the current water body management arrangement were expressed from different perspectives, through a process that allowed each interest group to discuss and then present their views. There was a specific focus on supporting the emergence of five different interest groups who were encouraged to hold informal meetings to discuss the challenges to community management and clarify their concerns.
2. A critical stage was a facilitated information gathering process through which villagers established the legal status of the local water body and developed relationships with key secondary stakeholders. Several weeks of discussions between community volunteers, the Land Office and the Upazila administration revealed that the water body was not 'closed' – that is, formally leased by the government. A cooperative from outside the village had previously asserted that it had rights over the water body, exploiting confusion over its status. This situation had been tacitly supported by key district level institutions to secure revenue from the 'leases', and enforced by the cooperative through the use of their *mastaan* – thugs – to collect fees from the village fishers. Lewins et al. summarize that, 'the process of information gathering and public debate has made secondary stakeholders at the Union Pareshad, Upazila and district level aware … of the common misinterpretation of the leasing system. It has also introduced villagers to the workings of those formal institutions that are supposed to represent their interests and opened up channels for continued dialogue. In this respect, the PAPD process has helped challenge the deliberately opaque access arrangements that prevent the poor from using their land and water entitlements' (2007: 49).
3. After three months of group discussions there followed a public meeting at which each group presented its concerns and priorities (Table 4.2). The issues presented by each group were discussed by the full meeting. A process of negotiation then resulted in a series of commitments undertaken by the community, intended to meet each of the groups' concerns. This public identification of how the different interest groups could have their legitimate concerns and needs met establishes locally recognized entitlements and, through the process of negotiation, set the stage for the formation of benefit sharing agreements.
4. The final stages of the process moved to a full action plan (including negotiated agreements, rules and organizations) and implementation. A further public meeting was used to promote knowledge sharing among the primary stakeholders at the village level and establish ways of overcoming potential barriers to implementation. The groups proposed an executive committee to ensure transparency in the management of the share system, but required a further round of facilitated discussion to ensure that women and the poor were adequately represented.

Table 4.2 Issues raised by different interest groups

Interest group	Community water body issue/concern
Rich landowners adjacent to water body	Concern over share of benefits, assumption being that share would be on basis of land holding. Practical Action concerned that this group could block change due to their vested interest in the current system
Poor – no land adjacent to site	Wary of the richer interests and strategies – so interested in minimizing costs to reduce the influence of the rich; concern over technical constraints on control of weeds
Poor – landless	Proposed equal share of benefits and the right to donate labour rather than capital; concerned about transparency in management and believed the rich could not manage alone; fingerling costs a major constraint
Full-time fishers	Felt they should have significant input on basis of knowledge, but concerned about their status leading to reduced input; may need compensation for fishing controls; weed control and fingerling costs recognized as technical constraints
Women	Motivated by the prospect of increased fish consumption and could contribute to drying stage; a full role would boost their skills for pond cultivation but concern that their input would be blocked by the wider community (never having had a labour role before)

Source: adapted from Lewins et al., 2007

Implementation commenced initially with 200 households becoming ordinary members of the water body management committee, each donating 300 Taka (US$4) under an arrangement that provides for equal profit sharing between the members. The popularity and potential of the plan resulted in a rapid rise to 400 households, or two-thirds of the village population. The management requirements for the final action plan, developed in the public meeting using the STEPS framework (Box 4.1), is summarized in Table 4.3.

Working to support communities in the charlands means engaging in a complex political environment in which multiple institutions overlap at the village level. These include the alliances within *gusthi* (kinship groups) and the trustee committee of each mosque (*masjid*, comprising influential community members). The *masjid* committees undertake management duties for the mosques, but also run informal dispute mediation and conciliation functions through *salish* sessions, which are the only mechanism for dealing with minor civil disputes. Overlaid on these informal village institutions is macro-scale politics and, more broadly, a context in which politics and disputes frequently escalate into violence.

As a process designed to encourage power sharing, PAPD has the potential to disrupt the established order and vested interests in this environment. The approach taken by Practical Action was seen by some as a challenge to the *gusthi* social structure in particular. Yet the potential for the process to release political and financial support for local initiatives was recognized by the kinship groups, drawing them in so that they would remain relevant. This

Table 4.3 Management requirements for the community action plan

	Management requirements
Social	Special provision to be provided to women and female-headed households Each household will have the opportunity to take part Villagers will share guarding duties
Technical	Fish sanctuaries will be used to boost local fish species Nets and bamboo enclosures to protect fish stock An additional flood-protected pond to be leased for fish seed production.
Financial	400 Taka membership is affordable for most and sufficient for the first year A monthly wage will be provided to the poorest guards Harvesting time according to market price – local traders will be used Committee to reinvest a share of profits (20 per cent) and redistribute the rest equally to participating households A provisional budget of 150,000 Taka will cover the first year costs
Institutional	The committee will operate a joint account with transactions requiring unanimous decision Expenditure and profits to be posted in the community building Banks, NGOs, Practical Action and the Department of Fisheries to be lobbied for extra support for fencing and stocking A professional fish seed merchant to be enlisted from outside the area

Source: adapted from Lewins et al., 2007

was a significant step, as changes in the social dynamics did emerge, evident in particular in the structure of the implementing committee and water body management committee. In focus group discussions community members observed that (Lewins et al., 2007: 57):

- Decision making had been more autocratic in the past, with 'influential leaders' taking the lead. With the formation of the committees, however, all participants recognized the need to establish roles for individuals and various factions such as the *gusthis*.
- Rather than one-off community wide meetings, separate meetings were being held in each part of the village and surrounding area. Marginalized voices were more likely to be honoured and the main *gusthi* became less dominant.
- Female representation had improved with four women formally represented on the new committees. As one women said: 'There has been a self-revolution! We have exposed weak points and strong points. There has been mental development and confidence building'.
- Before the PAPD committees were formed *salish* was the main means of resolving conflicts. This had merely worked to maintain the status quo – elites demonstrated power by using *salish* as a political tool. Once the new committees were operating with confidence, the coercion and violence associated with political issues became less of a concern.

The shift in confidence and emergence of rebalanced power relationships was most vividly illustrated by an exchange that occurred during the final stages of the action plan development. Lewins et al. report that (2007: 58):

> there was a major dispute over the allocation of future profits from the water body with the major landowner group demanding the greatest share. A meeting was held independently of Practical Action where the poorest group threatened to withdraw from the plan. The landowners ultimately agreed to an equal share of profits between the participants.

Following the completion of Practical Action's work, the independent decision making and actions of the villagers further illustrated the success of the new committees. In the subsequent year the water body management committee (WBMC) continued to meet regularly and the community made several modifications to the original plan to intensify fish production. The WBMC also went on to seek credit from another local NGO, allowing them to double their investment. In these ways, the local model of resource management demonstrated that it can sustain itself, develop new institutional linkages outside of the village, and undertake shared learning. Reflecting the goals of adaptive capacity building, Lewins et al. summarize that 'the process since the facilitated PAPD has been self-organising and there are indications that the broad management and planning approaches will continue to operate' (2007: 55).

Overall, PAPD addresses and illustrates many of the aspects of the first two components of supporting adaptive capacity – power sharing, and knowledge and information – discussed in chapter 3. It achieves this through a focus on the processes through which the poor can expand their opportunities to make livelihood changes. Power dynamics are acknowledged throughout, starting with the first PAPD step in which facilitators focus on expanding their knowledge of the local social and political context so that they can ensure that all groups are represented in the process. The remaining five steps are, in part, different approaches to uncover and address power relationships. For example, in the charlands the information gathering process revealed well-connected members of the cooperative to be using a combination of coercive, knowledge and political power to secure illegitimate control over the water body, supported by layers of administration that should have been upholding the rights and entitlements of the villagers. Further, by empowering representatives of the different interest groups present at the village level, power differentials were gradually eroded, securing, for example, reduced control of wealthy landowners, the negation of power residing in the *gusthi* and *salish* networks, and a fuller role for women. As participation in decision-making spread, the capabilities of poorer sections of the village community were expanded. Importantly for adaptive capacity, this took place through moves towards collective action that brought together the knowledge and information sources of a wide range of actors. Recognition was built for the significance of local knowledge and the partial nature of each group's understanding of the problems in the water body (for example, in the different social and technical issues identified in Table 4.2). This set the stage for the representatives and, ultimately, the wider community, to develop their knowledge through discussion and shared experiences. The implementation and subsequent revision of the water management scheme in particular demonstrates how shared learning can emerge from PAPD, with the partners in the co-management arrangement building their knowledge through their actions. Finally, Practical Action demonstrated how an NGO facilitator can bridge the different sections of the community and, where necessary, create the conditions for them to link with other secondary stakeholders.

While in the Bangladesh example PAPD was applied locally, the same principles can be used over wider areas at the regional or higher scales. Larger areas present a greater challenge as they may overlay several administrative units and will require particular care to incorporate the political support and concerns of powerful interests. *Practical Action's Facilitator's Guide* notes that this (Taha et al., 2010: 16):

> is particularly true in the urban setting where the concentration of formal government stakeholders is greatest and the economic stakes tend to be very high. Stakeholder analysis of regional institutions and their interests will help to focus the planning strategy on key stakeholders. The early stages of PAPD can provide evidence to decision-makers ... The task is then to quickly transfer the negotiation and discussion process to higher levels (civil society organisations and regional authorities etc.).

The challenges of facilitation therefore increase with scale, and the task becomes one of organizing and documenting formal and informal meetings between key actors. Negotiation will necessarily incorporate organizations that give regional representation to different interest groups, and the task will likely be longer and require greater flexibility than at the local scale, as shifting the established views and entrenched politics of larger groups may take more time. However, each situation is different and, as the Bangladesh example illustrates, even local scale politics can be complex and require a long process of facilitation to reach an agreement that is shared by all sides.

Pathways to power sharing

Participatory Action Plan Development is one among many approaches to sharing power and knowledge. For example, in describing work on building co-operative governance for integrated water resource management in South Africa, Colvin et al. refer to the 'development of an interactive approach to capacity building ... drawing from a broad portfolio of approaches variously described as social learning, social appraisal, or whole system development' (2008: 681). They refer to the use of tools as 'guiding metaphors' and design principles, reflecting the need to adopt a flexible approach to facilitation that responds to the local social and political context. The dialogue with stakeholders in the Mvoti sub-catchment illustrates this (Box 4.3), in which the 'U-process metaphor' was used to guide facilitation and generate understanding among participants as to where the discussions were leading (Figure 4.1). The U-process diagram is useful as it illustrates the stages of engagement in a process of power and knowledge sharing, summarizing the motivation behind the action-orientated stages in methodologies such as PAPD. It also proved to be an effective tool in Mvoti, where there was already agreement to enter a dialogue process and rapid understanding of the interdependencies between the stakeholders. As the facilitator's description of the process in Box 4.3 suggests, from this starting point the tool became useful in ensuring the success of a lengthy process, illustrating that a shared understanding of the consensus building process itself can be important in maintaining the commitment of participants.

Box 4.3 The Mvoti stakeholder dialogue

The Mvoti sub-catchment forms the northern border of the Mvoti to Umzimkulu Water Management Area and consists mainly of commercial timber in the upper reaches, subsistence agriculture dominating communal land inland around Maphumulo and sugar cane towards and along the coastal strip. With a predominantly rural population, irrigated agriculture and industry are the main water uses in the catchment, followed by afforestation and domestic use. Owing to the lack of significant storage in the catchment and the water requirements, which exceed the available resource, the catchment is considered to be stressed.

The idea of a dialogue process with stakeholders in the Mvoti sub-catchment developed out of the conversations held with the Department of Water Affairs and Forestry (DWAF) catchment management team. These discussions focused on the history of the team's work in the Mvoti over previous years and the difficulties they had experienced in supporting effective transformation of the two Irrigation Boards (IBs) in the catchment (Upper Mvoti Irrigation Board and Lower Mvoti Irrigation Board). It was agreed [to] seek to develop a stakeholder dialogue process to explore options for institutional arrangements in the Mvoti catchment, with a view to the transformation of the two Irrigation Boards and to seek wider benefit for communities across the catchment. At the same time, this would create a 'site of learning' for the DWAF catchment management team.

An early success of the dialogue was that stakeholders from the upper Mvoti (including the Upper Mvoti IB), the central Mvoti tribal communities, the lower Mvoti (including the Lower Mvoti IB) and the catchment-wide Mvoti Stakeholder Forum, agreed to work together through a dialogue process. Recognizing environmental, social and economic interdependencies across the catchment, they also agreed to work together to explore the option of a single institutional arrangement for the Mvoti. Another early agreement in the dialogue has been to involve a wider cross-section of stakeholders before different options are fully explored, resulting in a workshop for 50 to 60 stakeholders in March 2008. One key to the effectiveness of the process to date has been the shared understanding that building common understanding across diverse interests takes time if it is to succeed, and that this up-front investment of time could bring substantial benefits in the longer term. The 'U-process' metaphor has helped to frame this understanding, and to create a 'holding framework' for the dialogue as a whole.

Source: Colvin et al., 2008

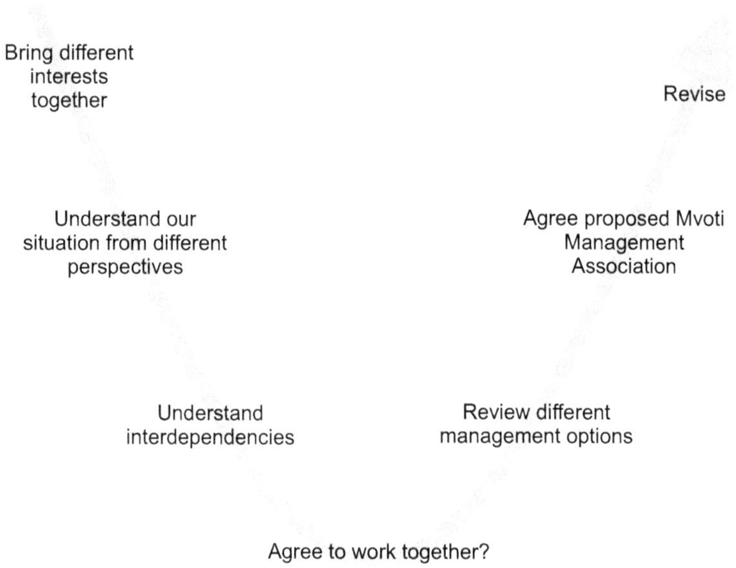

```
        Bring different
          interests
          together                                      Revise


        Understand our                           Agree proposed Mvoti
      situation from different                       Management
          perspectives                                Association



           Understand                             Review different
         interdependencies                       management options



                              Agree to work together?
```

Figure 4.1 The U-process metaphor applied to the Mvoti dialogue
Source: Colvin et al., 2008

More broadly, in their comprehensive guide to power sharing Borrini-Feyerabend et al. describe a wide variety of alternative methods and tools that have been successfully employed in achieving consensus in different natural resource management contexts. Through their comparison of examples, they emphasise the need for flexibility in applying these processes, but four steps are common: multiple stakeholders organizing and expressing their interests and concerns; negotiation of an agreement; setting up of a collaborative or co-management organization; and learning by doing through implementation (2007). The U-process metaphor in Figure 4.1 complements this synthesis by drawing attention to the development of knowledge through shared understandings of the different perspectives and interdependencies of stakeholders.

While the methods and tools for power sharing build relationships between local actors and those at wider scales, the emergence of effective power sharing arrangements may still depend on the conditions or restrictions that the wider policy environment place on local action. In these cases, attention may need to be paid to how specific policies are designed or (and) are being implemented. Depending on the prior knowledge of facilitators and primary stakeholders, these policy targets may be identified either prior to or during consensus building deliberations. In either case, two scenarios can result: either the need to work towards the integration of plans that result from power-sharing approaches into formal government strategy; or the need to influence broader, usually national level policy issues that impinge on community entitlements.

While the Bangladesh example above describes an intervention for local co-management that is independent of government, in other contexts it is desirable to integrate local planning into formal district or national delivery mechanisms so that adaptive capacity gains can be institutionalized and the community demands that result can be satisfied by the legitimate authority. Box 4.4 describes two very different approaches to securing support for local planning from government institutions, illustrating how the local political context determines the most appropriate course of action. While the specifics are different, in both examples it was necessary to build capacity with the communities and with members of the relevant institutions.

More broadly, Practical Action's experiences suggest that the capacity problems in 'meso' institutions (those layers of government that lie between the national and local level), particularly in rural areas distant from government ministries, can be a principal barrier to effective community-based action. While in many cases there remains a need to realign policies to work for the poor, the need to build the capacity of those with responsibility in these institutions to work towards consensus and power sharing with community

---

Box 4.4 DRR planning in different contexts: Nepal and Zimbabwe

In Nepal, it is government policy that each district should have a disaster committee responsible for establishing disaster plans and coordinating disaster risk reduction (DRR) activities. Practical Action has historically worked to build the capacity of at-risk communities and those that can provide vital services. In the preparation of community disaster management plans, Practical Action facilitated meetings between villages vulnerable to numerous hazards, with representatives from the district development committees, leaders of village committees, members of the local Red Cross, and experts on disasters and the local topology. In the process, participants were encouraged to recognize needs and options for support and the outputs were more comprehensive as a result. In addition, the presence of local and district authorities in the planning process achieved the integration of the plans into the formal procedures for public support, relying on Nepal's 'bottom up' approach for development planning that runs from the village development committees up to the national level. In two districts the strategies have succeeded in being included not only in village and district planning, but also in Nepal's national development plans.

In Zimbabwe a very different process was employed to achieve the same end. While in Nepal it is possible to undertake community based activities and then advocate integration at the district level, in Zimbabwe it is essential first to secure provincial government buy-in before work can take place on the ground. In this context, Practical Action undertook influencing at the provincial level, running workshops and trainings in participatory planning techniques, thereby building the capacities of government staff. At the same time, awareness raising amongst communities built demand for DRR. These twin strategies ultimately secured the release of teams from the district government to facilitate community-based planning.

While different in approach, both examples involve raising awareness in communities of the need to tackle disaster planning and of where the responsibility lies in government. By creating links between communities and the relevant layer of government, and building the capacities of those with responsibility, community-based disaster planning was achieved in very different contexts.

representatives is frequently essential. As has been identified elsewhere, the strategies for securing a role for communities range from strengthening the 'voice' of the poor, to strengthening the 'receptivity' of government (Goetz and Gaventa, 2001). Practical Action's experience is that work is often required at both ends of this spectrum.

The second scenario describes a situation in which larger-scale policy processes have the potential to derail or undermine the results of power sharing at the local level. If national policy emerges from a discussion amongst actors that exclude representatives of the poor, yet directly impacts on their livelihood choices, then efforts to support local adaptive capacity may demand engaging in influencing approaches to provide a voice for poor communities and to rebalance power relations. Many of the rights-based strategies discussed above apply here, in particular through the direct or strategic use of the law, where applicable, or by focusing on holding the state accountable through documenting or reporting on the impact of proposed policy changes.

In this latter category fall a class of activities that draw together local and national stakeholders and experts to develop new knowledge and relationships to achieve policy influence. As such, these approaches contain all the elements of power sharing, but rather than ultimately forming relationships for co-management, the objective is to provide a counterweight to the prevailing power holders in policy processes. Known as deliberative and inclusive processes (DIPs), these strategies would ideally include or be convened by the state as part of an integrated approach to policy development. Where this is not the case they can be used to good effect by non-state actors. Box 4.5 illustrates one approach to DIPs in relation to how the State of Andhra Pradesh (AP) attempted to transform farming and land use in line with 'Vision 2020' – the future of AP as imagined by the government. Convened by national and international civil society organizations, it represents a deliberate attempt to secure a space for the views of those affected by policy through a process that provides equal weight to different sources of knowledge.

There are many alternative DIP approaches, ranging from consensus conferences (in which lay people develop their understanding and views on technical issues) to visioning and future search exercises (in which a shared view of the future is developed). However, whichever approach is adopted, effectiveness depends on there being a clear relationship to the policy process in which influence is sought (for example, by ensuring that the DIP is considered in policy-making committees or includes key, influential individuals) and on how the process is designed, including attention to (Borrini-Feyerabend et al., 2007: 401):

- resources: accessibility of information, time, experts and materials to enable participants to engage effectively;
- methodology: clear and well-defined scope, procedures and expected outcomes;

Box 4.5 Prajateerpu: a citizens' jury/scenario workshop on food and farming futures in Andhra Pradesh, India

Prajateerpu is an exercise in deliberative democracy that involved marginal farmers and other citizens from all three regions of the State of Andhra Pradesh. The citizens' jury was made up of representatives of small and marginal farmers, small traders, food processors and consumers and indigenous people. Over two-thirds of jury members were women.

The jury members were presented with three different scenarios. Each was advocated by key proponents and opinion-formers who attempted to show the logic behind the scenario. It was up to the jury to decide which of the three policy scenarios most likely provided them with the best opportunities to enhance their livelihoods, food security and environment 20 years from now.

1. Vision 2020: This scenario was put forward by Andhra Pradesh's chief minister, backed by a World Bank loan. It proposes to consolidate small farms and rapidly increase mechanization and modernization of the agricultural sector. Production enhancing technologies such as genetic modification were expected to be introduced in farming and food processing, reducing the number of people on the land from 70 to 40 per cent by 2020.

2. An export based cash crop model of organic production: This was based on proposals from the International Federation of Organic Agriculture Movements (IFOAM) and the International Trade Centre (UNCTAD/WTO) and was based on environmentally friendly farming linked to national and international markets. This scenario was dependent on the demand of supermarkets in the North for a cheap supply of organic produce, complying with new eco-labelling standards.

3. Localized food systems: This scenario was based on increased self-reliance for rural communities, low external input agriculture, and the re-localization of food production and markets. It included long-distance trade only in goods that are surplus to local production or not produced locally.

The jury/scenario workshop process was overseen by an independent panel, a group of external observers drawn from a variety of interest groups. It was their role to ensure that each food future was presented in a fair and unprejudiced way, and that the process was trustworthy and not captured by any interest group.

The key conclusions reached by the jury represented their own vision of the desired future, including:
- food and farming for self-reliance and community control over resources;
- maintaining healthy soils, diverse crops, trees and livestock, and building on indigenous knowledge, practical skills and local institutions.

It also included an opposition to:
- the proposed reduction of those making their living from the land from 70 to 40 per cent in Andhra Pradesh;
- land consolidation in fewer hands and displacement of rural people;
- contract farming;
- labour-displacing mechanization;
- GM crops;
- loss of control over medicinal plants, including their export.

The Prajateerpu and subsequent events show how the poor and marginalized can be included in the policy process. The jury outcomes and citizen voice have encouraged more public deliberation and pluralism in the framing of policies on food and agriculture in Andhra Pradesh. The state government that championed Vision 2020 reforms was voted out of office in 2004. The largely rural electorate of Andhra Pradesh voted massively against a government it felt was neglecting farmers' needs, rural communities and their well-being.

Source: adapted from Pimbert and Wakeford, 2002; Pimbert and Wakeford, 2003; www.prajateerpu.org

- decision making: structured debate over underlying assumptions, how decisions are made and extent of public support;
- cost-effectiveness: investment in time and money is proportionate to the scale and importance of the decisions.

While DIPs are only one approach to policy influencing that development practitioners can engage with, they are significant to adaptive capacity because of the explicit focus on power and co-creation of knowledge. Even in cases where their eventual impact on a particular policy is marginal, these processes in and of themselves can support the adaptive capacity of the poor by developing their knowledge, relationships and influencing capacities, while demonstrating their capacity to contribute to policymaking to those who would normally overlook the voices of the poor.

Participatory technology development: experimentation and testing in practice

In chapter 3 participatory technology development (PTD) was introduced as an approach that focuses on people's ability to adapt technologies and innovate, embedding experimentation in the local context. PTD unpacks experimentation and testing through technology choice (the ability to make informed decisions over alternative technologies) and technical capacity (the skills and knowledge that individuals possess to engage in experimentation). In practice, however, these components overlap, as communities build their skills and knowledge through their experience with new and alternative technologies. The main activities that commonly constitute PTD reflect this, comprising elements that gradually expand choice and capacity:

- *exploring existing knowledge* – documenting and discussing local practice;
- *exposure visits* – building technology choice, sharing knowledge and stimulating relationships with external actors;
- *training* – informing, building technical capacity and cementing relationships with external actors;
- *experimentation and testing of technologies* – strengthening confidence and technical capacity through implementation and review (learning through action);
- *peer-to-peer dissemination* – sharing knowledge, building confidence and expanding project reach through community-led activities;
- *participatory research* – supporting the emergence of community-led agendas and participation in formal research (shared learning).

While not all interventions will include all these elements, PTD culminates in an increased confidence to experiment that grows through a process of testing new ideas and relationship building with external knowledge holders. In this way, community-based projects that foster experimentation also expand local knowledge, working with communities to build on their local norms and capacities. For example, in Sri Lanka, a Practical Action project focusing on rice production in saline affected areas built relationships between local farmer groups and a nearby research institution which had previously overlooked the needs of marginalized farmers (Box 4.6). The farmer-led research into alternative, traditional rice varieties demonstrates how local adaptive capacity – and in particular the confidence to experiment – can be fostered through the work of an international NGO with long-standing relationships to local farmers and access to actors at different scales. PTD was also a component of Practical Action's PAPD work in Bangladesh discussed above. By linking the char communities to agricultural research and support institutions, PTD was used strategically at an early stage in Practical Action's intervention, building confidence, cementing relationships and – crucially for PAPD – demonstrating the value of collective planning (Box 4.6).

This relationship between power sharing and experimentation and testing is explored in more detail below. First, in the following paragraphs the different elements of PTD are discussed in relation to a case study from Zimbabwe.

Box 4.6 Building confidence for experimentation

Participatory variety selection in Sri Lanka (Ensor and Weragoda, 2009)
The farmers conducted variety selection in their own fields. Initially 16 progressive farmers were involved in variety selection from 10 traditional varieties. Selections were made on the basis of scoring each variety on a range of qualitative indicators. Each variety was given a mark between 1 and 10, based on the farmer's preference, where 1 was the best available score. The criteria were established by the farmers and assessed plant height, duration of the crop, grain quality, grain colour, saline tolerance and the grain yield. The farmers planted up to 5 kg per variety in the saline-affected areas of their paddy fields and continuously observed the growth and changes in the plants up to harvesting. Farmers who cultivated saline-affected lands in the surrounding area were invited to the field during crop growth to make their own observations and discuss progress with the farmers involved in the trial. The participatory variety selection process enabled needs-based selection of varieties by the farmers and helped promote quicker adoption of useful varieties in the farming community.

Participatory technology development in Bangladesh (Lewins et al., 2007)
The PTD process implemented by Practical Action Bangladesh encourages technical choices by farmers and facilitates farmer innovations by helping to source technical and expert support through local representatives known as rural community extensionists. Villagers select a handful of farmers to receive extra training from Practical Action and government service providers in veterinary and cropping issues. The formation of PTD groups with a wide range of producers and stakeholders (including smallholders, full-time fishers and other landless, and women) helps create a platform for discussion of inter-related problems for char dwellers such as sources of seed and credit, market opportunities, linkage and relationships with service delivery organizations and personnel. The PTD initiatives were designed to enhance group dynamics and unity, bargaining power and interaction through discussion and planning in the villages. PTD represented an opportunity to instil a planning mindset and to raise the confidence of the poor and the most marginalized.

Case study: participatory technology development in Zimbabwe

In the 1990s, Practical Action (then known as ITDG – the Intermediate Technology Development Group) worked in the Chivi district of southern Zimbabwe using PTD to enhance food security in communities that worked the marginal *Chomuruvati* ('dark corner') lands. Chivi is characterized by general poverty and food insecurity in particular, with subsistence agriculture forming the mainstay of the household economy for most families. The project embodied three key principles that summarize its approach to PTD: building on local knowledge and skills; enabling participation in decision making as a fundamental step in increasing technical capacity and improving technology choice; and, strengthening local institutions, as essential to achieving participation. Written up as 'Beating hunger, the Chivi experience' this long-running project explored the different elements of PTD, as discussed in the following (Murwira et al., 2000: 59-73).

Exploring existing knowledge. Local knowledge is the starting point for PTD, reflected in the first principle underpinning the Chivi project. As a result, the intervention commenced by trying to understand what the community knew, using tools including semi-structured interviews and discussions with key informants, group discussion meetings and observations in crop fields and vegetable gardens. While this step provides the PTD facilitators with a fuller understanding of the context, it also opens a space for discussions around local technologies and the reasons for their decline or continued use. In Chivi, traditional pest controls had declined in favour of pesticides, which were generally perceived to be ineffective and in some cases were leaving dangerous residues. However, interviews revealed that farmers with knowledge of traditional alternatives, based on local herbs, were wary of talking about them for fear of ridicule in a community that had been exposed to more 'modern' technologies. Through the process of discussing these issues many farmers gained the confidence to start using their latent local knowledge of traditional pest controls, with the effect that technology choice was expanded.

Exposure visits. Engaging with external sources of knowledge becomes essential for issues that are beyond the scope of local knowledge, as is anticipated with climate change and, in the Chivi case, was the case with drought episodes that were unprecedented in the lifetime of the community members. Exposure visits provide the opportunity for community members to explore how others have addressed similar problems to their own. By visiting sources of knowledge outside the community, local representatives are better able to make an informed choice over a range of options, having developed knowledge rather than simply received information. There are also less constraints on selecting and rejecting options when the community themselves discuss and reach decisions than if an NGO project officer presents solutions to the community. As with power-sharing approaches, by taking the role of facilitator rather than knowledge holder, development practitioners are more likely to empower the community, enabling them to make and act on their own decisions. Murwira et al. describe the impact of the visits organized to research stations, NGO projects and other farms outside Chivi (2000: 67):

> After all the visits... the community met to get feedback from those who had gone. The report back sessions were very lively. Some farmers drew on the ground what they had seen and how it worked. Those who had had the opportunity to see comparative situations of before and after the technologies were adopted were able to share with the others how significant the difference was ... each technique was described and weighed in great detail; how easily adaptable it was, how much it cost, how heavy, how much input was needed materially or financially.

Summarizing the value of exposure visits, one village leader commented: 'This man showed us these pits that he had dug to 'hold the water' in his fields. It looked so simple and I said, how come we never heard of such a simple thing before? That is why in Karanga it is said *kusaziva hufa* (lacking knowledge is as good as being dead)' (ibid.: 65).

Exposure visits also start a relationship building process between community members and external actors that is vital if adaptive capacity is to include the ability to experiment with and test new ideas. Enabling a community to access and have the confidence to work with alternative knowledge is a key contribution of PTD to adaptive capacity. However, the representatives that are selected for visits will determine the effectiveness of the process, and must include a cross-section of the community. In the Chivi case this meant ensuring gender equity (by covering both field cropping and vegetable gardening issues), including both literate and non-literate community members, and local leaders as well as non-leaders from the villages. The facilitators encouraged the community to decide on who was to participate by reaching a consensus in planning meetings.

Training. Following the exposure visits, facilitators worked with the community to reach a consensus on the most promising options for implementation. Training was arranged in the villages which, in the first instance, was run by individuals brought in by Practical Action from the research centres or projects that had been visited. Later in the project the villagers themselves provided exposure visits, sharing their experiences and giving training for farmers from elsewhere in Chivi district.

Experimentation, testing and dissemination of technologies. Having received training, local farmers began to test and adapt technologies in their own fields. Farmers made many modifications on their own initiative, while some initiated comparative trials in which they adopted the technologies in different ways across their plots. Practical Action facilitated reviews and knowledge sharing sessions during and at the end of each season in which the farmers exchanged innovations and their analysis of their experiences. Through this process, technical capacity, confidence and the ability to make changes were enhanced: contemporary observers reflected that there was 'an explosion of experimentation by local farmers, turning fields into plots for testing new ideas for soil

> **Box 4.7 Sharing knowledge: seed fairs in Chivi**
>
> The idea of the seed fairs arose out of the problem of [finding] suitable dry land crop varieties for the climate in Chivi. The objective of the fair was to revive local seed varieties, share information on these and acknowledge that these crops do thrive in Chivi. At the seed fair, which is like a trade fair, farmers display local seeds, for example, mhunga, millet, rapoko and so on. The farmer with the best-looking varieties is awarded a prize. Each year these fairs get bigger and better; they are planned and organized by the community members, who set the judging criteria themselves and also contribute the prize money. This has boosted farmers' confidence and increased the sense of ownership over their own activities. In the past, shows and field days were judged according to criteria set by extension workers of outsiders, which were not always the same as the farmers.
>
> *Source:* Murwira et al., 2000: 69

and water conservation and for planting techniques' (Mulvany et al., 1995, quoted in Murwira el al., 2000: 73). As a consequence, farmers also began to ask for information on alternative crop varieties that would be suitable for their dry land farms. Practical Action facilitated a process of obtaining and, via local farmers' groups, the dissemination of seeds that the farmers then experimented with in their fields. These trials further increased the confidence and skills of the farmers to test and adapt technologies that they had no previous experience of working with. Throughout this period of experimentation and testing, dissemination occurred organically through friends and neighbours observing progress and based on their trust in the analysis of their fellow farmers. However, Practical Action also sought to stimulate knowledge sharing and technology dissemination. This was done through events such as exposure visits and training run by the newly trained farmers; the attendance of non-participating farmers at the review sessions; and the instigation of competitions and 'field days'. As Box 4.7 explains, Practical Action introduced community-led seed fairs that were particularly effective in demonstrating the potential of the new technologies and stimulating knowledge exchange among farmers from the surrounding areas.

Participatory research. The final aspect of the Chivi project involved attempts to influence those in formal research institutions to work more directly with farmers and to shift their research focus onto issues of significance to the local farmers. As noted above, formal research institutions frequently overlook the needs of poor farmers, with conventional practice placing the scientist at the centre, determining both the area of focus and the criteria for success (Scoones and Thomson, 2010). However, the relationship building that took place during exposure visits and training sessions provided the starting point for changing the perspective of researchers and, through Practical Action's facilitation, researchers were encouraged to work more closely with farmers and instigate formal trials in their fields. The result was a process of shared learning in which both the researchers and farmers gained new understandings: following poor performance of rigidly imposed trials, the farmers 'persuaded the researchers to allow them to conduct the trials ... using their own knowledge of when to plant and so on, with far more successful results' (Murwira, 2000: 71). As experts in their own complex environment, the farmers also employed much more detailed criteria for success than the researchers, focusing on multiple attributes including reliability, pest resistance, storage and ease of processing, in place of the single over-riding concern of the researchers for drought tolerance.

Power-sharing approaches and PTD

Through the emphasis on participation, knowledge sharing is central to the activities that characterize a PTD intervention. This occurs in several phases, first through exposure visits in which the community members have the opportunity to discuss unfamiliar practices with those experienced in their use, then through the dissemination and consensus discussions with the rest of the community, and finally through experimenting with technologies and sharing learning with friends and neighbours. As noted above (Box 4.6), in Bangladesh, these phases were used instrumentally to support the emergence of PAPD, in particular by demonstrating the value of collective planning. For the purposes of supporting adaptive capacity, it is significant that the reverse is also true – that the power sharing strategies developed by working through PAPD can support experimentation and testing through PTD. Through their structured approach, PAPD and similar methodologies provide a framework that can incorporate the components of PTD, embedding the knowledge sharing components of experimentation and testing into the process of building power sharing relationships, as Table 4.4 suggests. The advantage lies in enhancing the 'P' in PTD through the adoption of PAPD's rigorous approach to participation and relationship building, while integration also provides a way for development actors to bind together the three dimensions of adaptive capacity: power sharing, knowledge and information, and experimentation and testing. The result is a development process that builds the relationships between the primary and secondary stakeholders necessary to create changes in a particular social, political and environmental context, within which the ability to experiment and test is supported through enhanced technology choice and technical capacities.

Table 4.4 Potential synergies between PAPD and PTD

PAPD step	PTD activity
Preparation and problem census	Exploring local knowledge and establishing local priorities for alternative technologies
Information gathering	Exposure visits
Analysis of solutions and public feedback	Community feedback and building consensus on technology options
Action plan development and implementation	Training, experimenting and knowledge sharing

PTD's relationship to methodologies such as PAPD goes a step further when attention is turned to opening up formal research to participation. Here, there is an explicit demand for power sharing between the scientific establishment and the communities with needs waiting to be met. Alongside livelihood focused agendas such as those of agricultural research stations, this also includes meteorological institutions, whose products need to respond and be accessible to vulnerable communities (Box 3.3). Through its focus on participation, the logic of PTD demands that control over research agendas is opened up and the power to make decisions is shared. The arguments in favour of power sharing developed in chapters 2 and 3 hold equally here: in a complex system, the perspective of one group of stakeholders is unlikely to be sufficient to fully understand the nature of the challenges being faced. Out of right and out of necessity, community participation is essential in determining the future direction of science and technology research and development.

The strategies for power sharing described above can therefore be applied to participation in research. First, consensus building approaches such as PAPD can be applied directly to technology gaps or challenges identified by communities by involving key research institute personnel as stakeholders from an early stage in the process. In PAPD, these individuals would be identified as secondary stakeholders during the information gathering stage, and would need to be closely integrated into the subsequent analysis and action plan development in order to build relationships and develop a shared research agenda. In reality, flexibility and responsiveness to the local context will likely be necessary, with the different stakeholder groups needing to develop relationships from the outset (as suggested by the U-process in Figure 4.1) in order to develop trust and a shared commitment to the process.

The second power sharing strategy is the use of deliberative inclusive processes (DIPs) to influence the direction of science and research policy. The use of citizens' juries, consensus conferences or future search exercises is needed to ensure the different interests and priorities of marginalized communities are represented in the research and funding choices made in the natural and social sciences. DIPs can be used by those supporting adaptive capacity as a mechanism to promote power sharing in research, and thus experimentation and testing that is undertaken by or relevant to the interests of the poor, by focusing on three key areas (adapted from Pimbert, 2007: 16):

- Reorganizing conventional scientific and technological research to encourage participatory knowledge creation and technological developments that combine the strengths of poor communities and scientists in the search for locally adapted solutions. Effective and interdisciplinary partnerships are needed to link natural and social sciences with indigenous knowledge to address needs and problems in specific local settings that are typically marked by complex and dynamic change. An important goal here is to ensure that knowledge, policies and technologies are tailored to the diversity of human needs and the situations

in which they are to be used. This must be on the basis of an inclusive process in which the means and ends of research and development are primarily shaped by and for citizens through conscious deliberation and negotiation.
- Opening up decision-making bodies and governance structures of research and development organizations to allow a wider representation of different actors and greater transparency, equity and accountability in budget allocation and decisions on research and development priorities. They are immensely powerful in that they broadly decide which policies and technologies will ultimately be developed, why, how and for whom. And yet the governance of science and technological research and development is presently largely dominated by men who are increasingly distant from rural realities and moving closer to corporations.
- Ensuring knowledge and innovations remain accessible to all as a basic condition for economic democracy and the exercise of human rights, including the right to participation. Decisions to issue patents on knowledge embodied in products and processes (seeds, software etc.) and national intellectual property rights legislation require more comprehensive public framing of laws and policies based on deliberative and inclusive models of direct democracy.

Experimentation and testing are critical to adaptive capacity because the uncertainty of climate change demands that communities have access to and the ability to work with novelty. Supporting this means not only enhancing choice and capacities through PTD, but also enabling communities and researchers to work together through attention to local-scale power and knowledge sharing relationships and the larger-scale constraints, incentives and interests that govern the direction of science research.

Adapting development practice

This chapter has presented an overview of approaches to development practice that offer lessons and tools for those seeking to support adaptive capacity. These demonstrate that the challenge of supporting adaptive capacity is not one that is entirely new to development practitioners: there are many experiences and examples of practice that can be drawn on and woven into development work. Less important than the particular tool or methodology is the necessity to respond to the local context in addressing the issues and principles that define support for adaptive capacity – to fulfil the requirements of chapter 3 for the reasons discussed in chapter 2. Building support for adaptive capacity is itself a process of fostering the confidence and capacities of local people, better suited to the goals of a long-term programme of work than a single project cycle. Within programmes, separate projects can develop the different dimensions of support while responding to local needs. For example, a project that accesses and tests new technologies can provide an entry point to working with communities, while also laying the foundations for subsequent power sharing that draws in knowledge and stakeholders based on the lessons learnt and challenges identified.

The examples of PAPD and PTD presented here demonstrate that a coherent approach to supporting adaptive capacity can be developed by NGOs through their programme work. The common focus on shared learning defines processes that require different interest groups to share knowledge, respond to information and test ideas through concrete actions. As described in chapter 3, knowledge and information feed into power sharing relationships and emerge as collaborative actions – experiments and tests – that apply new understandings and produce learning, linking the elements of support for adaptive capacity. However, as PTD emphasizes, information and experiences from experiments and tests outside the local context are a vital component in supplementing local knowledge and stimulating contextualized experimenting and learning. And as the Bangladesh experience of PAPD illustrates, engagement with power sharing arrangements is about developing self-sustaining learning systems that respond to new knowledge or changing circumstance through the agency of actors that come together for a common purpose.

In short, a closed system is not the end result that is being sought. Rather, adaptive capacity results when the interests of the poorest are represented in systems that draw in resources, revise plans in light of learning and send out demands or make claims on external actors (as in Bangladesh where the governance committees called on new financial resources to expand and improve their fisheries). Figure 4.2 represents this overall goal. The purpose of the three dimensions of support for adaptive capacity is not that they be seen

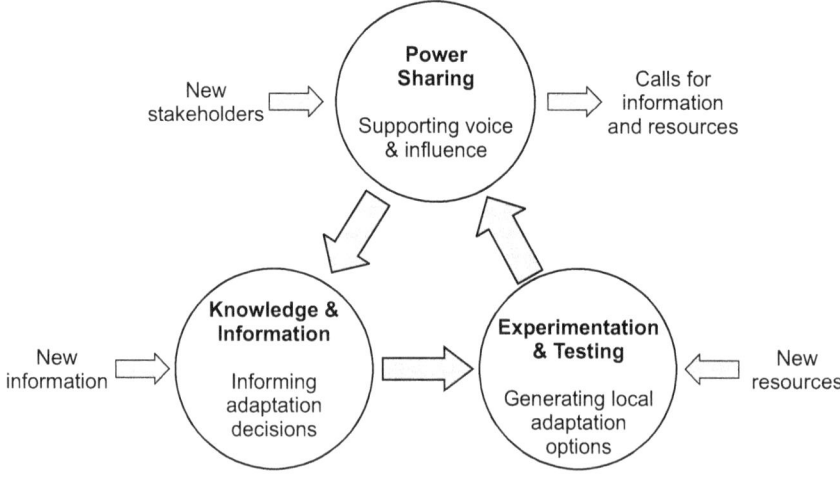

Figure 4.2 Towards self-sustaining support for adaptive capacity

as an end in themselves, but rather that they set in place the conditions for emergent knowledge, actions and power sharing arrangements.

As the lessons of rights-based approaches suggest, this understanding of supporting adaptive capacity is about transforming relationships and empowering the poor and marginalized through facilitation and support. This is about capacity building, but not in the co-opted neo-liberal sense of training for productive capacity, but in the original sense of emancipatory politics. As Deborah Eade describes, 'the role of the engaged outsider is to support the capacity of local people to determine their own values and priorities, to organise themselves to act upon and sustain these for the common good' (2010: 206). But working in this mode is a responsibility that challenges NGOs to think and work differently. In Chivi, for example, all community members were to a greater of lesser extent exposed to 'training for transformation', designed to (Practical Action, 2011: 2):

> empower individuals, leaders and different socio-economic groups by raising their critical consciousness, stimulating and encouraging them to actively participate and take control of issues that affect their lives. The training is based on Paulo Freire's method ... assisting people to understand the context of poverty and the need for alternatives to the dominant development approaches. It also shapes and influences the why and how of development.

Understanding poor communities as rights-holders locates NGOs as duty-bearers as much as it does the institutions of government. This is a challenging transformation that demands time and resources be channelled into equalizing the power relationship between NGOs and the communities they work with, through the search for processes that institutionalize 'downwards-accountability' and methods that empower the poorest. The length and depth of engagement with communities is longer and deeper than that required to justify interventions to donors or required by a mode of development practice that is defined by spending money through ever-greater numbers of projects. Alongside the search for co-management and shared learning, then, sit the less comfortable ideas of 'co-development' and 'shared risk' in which development organizations and communities work, learn and change together (Eade, 2010: 212).

CHAPTER 5
Uncertain futures

This book addresses how poor and marginalized communities can respond to climate change. The focus on adaptive capacity derives from a well-established fact: being certain about the existence of climate change is not the same as being certain about its impacts. It is this uncertainty that defines the challenge of adaptation, presenting development actors with a new challenge. How can communities be supported to meet the emerging reality of climate change and secure well-being in their uncertain futures? As Chapter 1 identified, the starting point is to seek a balance between actions that support the ongoing ability to change and those that respond to current challenges. Immediate needs, however, can no longer be seen as ends in themselves.

Chapter 2 adopted a resilience lens to bring into focus two issues of significance for development actors seeking to support this ability to change. First, adaptive capacity was defined in relation to resilience. This perspective is essential to understanding adaptive capacity, as it identifies uncertainty and change as normal, and rejects the idea that communities and the natural resources on which they depend can exist in a balanced, fixed equilibrium. Adaptive capacity is therefore employed by actors to make changes that increase the prospects of a community maintaining its well-being in the face of external disturbances such as climate change. As such, it encapsulates the ability to increase the resilience of their existing or transformed livelihoods. Second, learning derived from the study of social-ecological systems was introduced to understand how adaptive capacity can be supported. The complexity of real systems determines the nature of the challenge: to better understand how climate change and proposed adaptations play out, adaptive capacity must integrate knowledge and information from multiple stakeholders at different scales. At the same time, it must secure the ability to influence policies and regulations in favour of local action, while ensuring access to experiments in new ways of working or living, preventing livelihoods from becoming fixed and securing options for the future.

Chapter 3 translated these insights into three interconnected dimensions of development support for adaptive capacity: seeking power sharing arrangements that expand networks, voice and influence; securing the sources of and processes that give rise to knowledge and information; and ensuring the availability of experimenting and testing of locally appropriate adaptation options. These dimensions place power and knowledge sharing in the centre of efforts to secure opportunities for change. The experiences of working with rights-based approaches, discussed in chapter 4, suggest that this has important implications for development work: it must be transformative and political, with a focus on entitlements and accountability. As such, the law may have a role to play in securing opportunities for adaptation. Yet this is not entirely new territory for development practice, and as chapter 4 illustrates there are ample tools and case studies that can assist those seeking to support the emergence of adaptive capacity.

Ultimately this understanding means that adaptive capacity cannot be achieved through a particular development intervention. Rather, it points to supporting communities in their ongoing processes of building and rebuilding relationships, knowledge and capacities, in ways that enable them to better meet the challenges of climate change. The implications of climate change are, however, only gradually becoming clear. As international negotiations continue to fail to make real progress towards meaningful cuts in greenhouse gas emissions, this concluding chapter revisits the question of uncertainty and asks what a failure to limit climate change to 2°C average global warming means for this understanding of adaptive capacity.

The first observation is to note that even with the prospect of increased emissions, there are fixed points in the discussion of climate change (Stafford Smith et al., 2011). Uncertainty remains a fact of life. The sources of climate uncertainty identified in Box 1.1 will continue to beset climate projections in the future and, if anything, are being compounded as scientific understanding reveals new unpredictable or unclear climate processes. Despite this, it is important to recognize that some impacts of climate change will remain reasonably foreseeable, such as on global average temperatures, sea-levels and ocean acidification, all of which will continue to increase and whose minimum changes are reasonably certain even in the longer term. On the other hand, precipitation will remain unpredictable in terms of overall rainfall and distribution, particularly at the sub-national scale. This situation is not improving: recent observations suggest that existing models may be underestimating the potential for future changes in extreme precipitation (Min et al., 2011).

This pattern of future uncertainty, in which the more or less unpredictable impacts of climate change bring new disturbances to complex human systems, provides the context into which the consequences of supporting adaptive capacity will be played out. The remainder of this chapter builds on this to discuss how support for adaptive capacity relates to different types of futures communities may experience as climate change continues to take hold. The most straightforward of these is livelihood transitions. Gradual changes to livelihoods may be possible (and necessary) in many circumstances, through which communities are able to adjust to emerging climate change. However, where the consequences of 2°C or more of global warming bite harder and the impacts of climate change are more severe, livelihood transitions may not be enough. Rapid changes to natural support systems may pave the way for livelihood transformations and the emergence of conflicts as the available resources change. As the next two sections suggest, these possible futures underline the need to focus development support on adaptive capacity and towards empowered communities with expanded opportunities.

Livelihood transitions and transformations

The need to make incremental changes to livelihoods is likely to be the first implication of climate change for most poor and marginalized communities. Both gradual warming and increasingly erratic rainfall are already bringing new challenges to communities, necessitating livelihood transitions in which the same overall functions are achieved but with increased resilience. Examples include the introduction of disaster risk reduction measures such as flood early warning systems, or planting new varieties of crops to maintain local food production in a drying climate. The majority of practitioner attention on adaptation to date focuses on transitions, usually divided between actions in response to increasing disaster risk or the failure of existing livelihoods (for example, Ensor and Berger, 2009a; Pettengell, 2010; UNDP, 2010b). In many circumstances 'no-regrets' adaptations are sought, in which the benefit is not dependent on a particular climate future. These approaches frequently also improve the ability of the community to absorb disturbances, such as through interventions that build soil fertility (leading to enhanced crop health and regulation of moisture) or develop more effective rainwater capture mechanisms (securing water access in the face of erratic rainfall).

The decisions taken in many livelihood transitions are made with confidence because they respond to emerging climate conditions and have short lifetimes (where 'lifetime' refers to the lead time needed to bring it into effect plus the period that it remains in effect). Decisions as to which existing crop variety to plant, for example, have short lifetimes as a different choice can be made every year. By contrast, longer lifetime adaptations, such as construction projects or the development of new cultivars, have implications for those concerned that last for much longer periods (Stafford Smith et al., 2011). The confidence in short lifetime decisions thus derives from the attention to existing problems (the challenge is understood) and their flexibility in the face of future climate change (different actions can be taken if the problems change). However, short lifetime decisions still need to consider short-term uncertainty, such as where rainfall variability is increasing. Here, uncertainty may motivate the selection of a diversity of varieties over one whose yield is optimal over a narrow set of conditions. The complexity issues identified in chapter 2 also remain even if decisions are flexible in the face of future change. Multi-stakeholder approaches that expand the pool of experience in decision making remain essential due to the potential for local adaptation actions to have unforeseen consequences, or impacts on actors at other scales. Similarly, there may be local consequences of changes that are being made elsewhere, or the need to secure new resources, knowledge, information or freedoms to support local changes. In each case, processes that draw in all relevant stakeholders can support more effective decision making by developing a shared understanding between stakeholders of the problems being faced.

Equally, it remains essential to pursue adaptation actions through approaches that support future changes. The nature of climate change (and of complex systems) is that today's conditions are a foretaste of many decades of change rather than of a new equilibrium climate. Supporting the emergence of adaptive capacity – in the form of a sustainable realignment of power relations, access to multiple sources of knowledge and information, and links to and capacities in experimentation and testing – remains a critical goal of adaptation work in developing countries even when straightforward incremental livelihood transitions may be achievable. Actions that focus exclusively on immediate term challenges inevitably delay progress towards adaptive capacity, leaving communities exposed to future climate change. The necessity of bringing adaptive capacity to the centre is amplified by the increasing prospect of 4°C or more of global warming, now considered likely by the end of the century under a roughly business-as-usual emission scenario, with 2°C warming being reached between 2045 and 2060 (Betts et al., 2011).

This reality – itself a consequence of a failure of political will on emissions – underlines the urgency of moving to an adaptation mindset that is focused on continuous and largely unpredictable change. The consequences of 4°C warming are potentially devastating, including significant and rapid sea-level rise, meaning the likely displacement of tens of thousands, if not tens of millions of people; increased water stress, driven in some areas principally by climate change rather than population growth; and reduced crop yields, with rain-fed crop failures potentially occurring one year in two in southern Africa (New et al., 2011). The overall implication is that the world – and in particular the developing world – can expect more rapid and amplified impacts compared to those assumed under a 2°C scenario that there is 'little to no chance' of achieving (Anderson and Bows, 2011: 369). This translates into an urgent necessity to work with poor and marginalized communities to ensure that they are not overwhelmed by the need for frequent livelihood transitions, and are prepared for the possibilities of increased resource competition and livelihood transformation.

The prospect of 4°C warming reinforces the case for enabling communities' access to climate information and increases the pressure on decisions that have lifetimes beyond a few years. Under a 4°C scenario, uncertainty increases more rapidly with time, eroding confidence in future climate conditions more rapidly than in a world limited to 2°C warming. The prospect of successive incremental changes cascading into a process of near-continuous change means that even short lifetime decisions need to be informed by the context of climate change. Longer lifetime changes, such as community or farm infrastructure projects, need to cope with the reality of rapid and unexpected change, demanding local decision making processes that not only integrate multiple perspectives (including climate scientists) but also account for multiple plausible futures. The implication is that for longer lifetime decisions and for raising awareness of climate change, scenario-based exercises will be needed to help all stakeholders build a picture of the different ways in which the future may unfold. Through facilitation that helps stakeholders translate climate uncertainty into potential local impacts, alternative futures can be envisaged. In this way, shared decision making can account for the consequences of different actions under different conditions, guiding actions towards those that are able to build resilience in the face of uncertainty (Stafford Smith, 2011). Livelihood transformations, in which new ways of living are adopted to capitalize on or cope with the changing environment, are decisions that have long-lasting consequences. As such, uncertainty plays a major role, even if transformations are prompted by current conditions that have rendered existing livelihoods untenable. Ultimately, however, livelihood transformation requires the same preconditions as transition: informed, shared decision making that minimizes risks from uncertainties, maximizes knowledge, and learns from examples of and emerging experience in alternative modes of living.

Living in the future

For many of the poorest and most marginalized, living with 4°C or more of warming will mean living with profound environmental change and potentially devastating social impacts. Conflict has received increasing attention in climate change debates, often posited as the worst-case scenario of increasing climate impacts. The presumed drivers focus on increasingly scarce natural resources driven by sea-level rise, extreme events and gradual desertification or water stress, all of which are projected to be more severe in a 4°C world. Here, conflict is usually shorthand for 'violent conflict', meaning that disputes escalate into violence when competing parties resort to physical force to achieve their objectives. Livelihood transformations may themselves ignite disputes as new winners and losers are created, while, among other impacts, food and water scarcity and enforced migration could overwhelm the resilience – and well-being – of thousands if not millions of people. This section reflects on the sufficiency of adaptive capacity, as presented in the previous chapters, to build resilience in face of these ecological and social pressures.

The starting point for any discussion of climate change and conflict must be that the relationship between them is neither well understood, nor straightforward. For example, while resource scarcity is assumed to be a trigger for increased competition and an escalation in the likelihood of violence, research documenting two decades of social conflict in Africa suggests that unusually wet years, when resources were abundant, saw a greater increase in violence than dry years (CCAPS, 2011; see Box 5.1). While there is also an overall correlation between extreme weather events and the emergence of violence, this closer analysis suggests that it is resource competition, not scarcity per se, that is a significant driver of conflict.

More broadly, the relationship between conflict and climate is indirect and complex. The consequences of natural disasters and food or water insecurity are played out in different social and political environments. It is how disputes, shortages and social disruption are handled in society that determines whether they are a trigger for conflict, with violence more likely in

> **Box 5.1 Cattle raiding in East Africa**
>
> Many communities, particularly in semi-arid regions of Africa, depend on raising cattle for their livelihoods, and depend on the natural environment for their well-being. Poor rainfall can spell disaster as herds are depleted through thirst and inadequate grazing. Yet, pastoral communities in countries such as Kenya, Uganda, Ethiopia and Sudan are often at the margins of society, with negligible government presence in remote areas to arbitrate disputes. Consequently, violence is often used to settle conflicts, and cattle raiding – while a long-standing practice – has become especially deadly in recent years.
>
> In 2009 and 2010 in Southern Sudan, for example, various ethnic groups including the Dinka, Nuer, Murle, Mundari, and Shilluk tribes clashed over cattle, water, and grazing rights, resulting in hundreds of reported deaths. Contrary to popular belief, however, new research shows that cattle raids in East Africa are more likely to occur in wet years rather than during periods of drought and scarcity. When rainfall is abundant, cattle have more access to water and grass for grazing, making them fatter and more profitable to steal; whereas drought stricken herds are not as economically valuable. Moreover, dense vegetation makes it easier for raiding parties to conduct ambushes and to safely escape. Therefore, governments in the region and international agencies must pay greater attention to not only helping pastoralists cope with drought, but also to shoring up alternate, peaceful means of settling disputes.
>
> *Source:* CCAPS, 2011

the context of poverty, political marginalization and weak governance (Smith and Vivekananda, 2009). Moreover, the emergence or threat of violence can restrict access to some resources, placing greater pressure on others – such as where pasture is abandoned due to armed conflict, leading to overgrazing in the remaining lands. Similarly, migration in the face of climate risks or disasters can create population pressures in areas perceived to be safe. These feedbacks can generate downward spirals of resource competition and conflict that trigger or sustain violence (Buhaug et al., 2008; Saferworld, 2010).

Appreciation of this complexity should guard against overconfident assessments of the likelihood of violent conflict emerging as a consequence of climate change or responses to climate change. Few conflicts have a single cause. At the same time, while there are difficulties in anticipating violence, climate change remains a potential driver. Conflict analysis is helpful as it directs attention towards the different factors that mediate climate impacts in a particular context: natural resource management, local livelihood options, and, ultimately, local security and justice mechanisms. In many contexts, these factors overlap, with local livelihood options closely related to natural resource management, and dispute resolution occurring in natural resource management mechanisms, such as in pastoralist grazing committees (Saferworld, 2010).

This analysis has three important implications for adaptation. First, unresolved conflict can have consequences for adaptation. To take the example above, reduced access to pasture due to conflict can shut down pastoralist migration routes that would otherwise have been relied on to cope with climate-induced water or forage shortages. The vulnerability and natural resource pressure that results adds to the existing sense of insecurity, potentially fuelling a further cycle of conflict. Second, adaptation can itself have consequences for conflict. Adaptation actions that support or accept existing power relations provide preferential resource access to one group at the expense of another, further marginalizing the poor, exacerbating inequalities and setting the scene for dispute and conflict. The third implication is, therefore, that navigating existing disputes and systems of power and equity are critical in accounting for conflict when planning adaptation actions. Unlike directly implementing adaptation measures, supporting adaptive capacity through the power and knowledge sharing approaches outlined in the preceding chapters achieves this, enabling action on climate impacts while ameliorating conflict. As the peace-building NGO International Alert puts it (Smith and Vivekananda, 2009: 4):

> The double-headed problem of climate change and violent conflict thus has a unified solution – peacebuilding and adaptation are effectively the same kind of activity, involving the same kinds of methods of dialogue and social engagement, requiring from governments the same values of inclusivity and transparency. … A society that can develop adaptive strategies for climate change in this way is well equipped to avoid armed conflict.

Consensus building approaches, such as participatory action plan development (PAPD, chapter 4), are fundamentally conflict resolution strategies: they are structured mechanisms for bringing together stakeholders with different interests. While these approaches are not dependent on pre-existing conflict, they have their roots in efforts to mediate between parties with entrenched and divergent positions (Lewins et al., 2007). By establishing new communication networks and decision-making opportunities, they not only help to transcend conflicts but also help poor communities to explore new livelihood options – thereby fulfilling a crucial role in conflict resolution and adaptation.

Existing security and justice mechanisms also have an important role in successful support for adaptive capacity. Practical Action's experience of working through consensus building approaches has highlighted the significance of traditional institutions that have locally recognized sources of authority. For example, in addressing cattle rustling and banditry in northern Kenya, PAPD was structured around the indigenous conflict management system centred on elders or chiefs' councils, while in Darfur (Sudan), integrating the *Ajaweed* has been key to ensuring decision making and planning respects local norms (Box 5.2). Similarly, by working with the *salish*, the PAPD committees in Bangladesh reduced violence by moderating its influence, but also reinforced the legitimacy of the PAPD process itself (chapter 4).

The treatment of migration is in many ways similar to the handling of conflict in climate change debates. As with conflict, it has emerged as a potent

> **Box 5.2 Informal institutions for conflict resolution and local decision-making**
>
> In Sudan, Ajaweed is the traditional system for conflict resolution in rural areas and is recognized by both government and civil society. Ajaweed operates in all rural Sudan but is particularly important where seasonal migration brings low-level conflict. Ajaweed uses a recognized group of wise people in each village cluster who are entrusted by the community to resolve disputes and help address grievances. At state level there is an Ajaweed Committee composed of tribal leaders and other respected individuals that can address larger and more complex types of conflicts. The Ajaweed Committee has even been able to resolve local issues within the Darfur conflict.
>
> In northern Kenya, the elders have a number of sources of authority that make them effective: they control access to resources and to marital rights; they have access to networks that go beyond the clan boundaries, ethnic identity and generations; and they possess supernatural powers reinforced by superstitions and witchcraft. Their function as a court is to interpret evidence, impose judgements and oversee reconciliation. Parties typically do not address each other, eliminating direct confrontation, and interruptions are not allowed while parties state their case. Following cross-examination, visits to dispute scenes, private consultations and reviewing of previous cases, the elders then use their judgement and position of moral ascendency to find an acceptable solution.

topic due to the potential for large-scale population movements in response to livelihood-threatening impacts – but there is little evidence to support a simplistic narrative of impending catastrophe. Despite this, a significant and sometime highly charged discussion that centres on the prospect of up to 200 million environmental refugees has developed (Hartmann, 2010). Yet the effect of climate impacts depends first on how climate change manifests itself in a particular location, and second on how people in that location respond. As with conflict, this leads to a 'double uncertainty' in which social and environmental factors overlap to determine the ultimate impact (Gemenne, 2011: 183), dependent 'crucially on country-specific and contextual factors' (Buhaug et al., 2008: 2). The most alarming figures for potential climate migrants rely on an estimate of the number of people at risk, rather than of those likely to move, and as such have received widespread criticism in the face of an emerging consensus on the importance of 'multiple and overlapping causes of most migration flows, including economic, social and political factors' (Tacoli, 2009: 107).

The empirical evidence that exists linking environmental change and migration reflects this much more complex picture. In some cases, such as sea-level rise, there can be a reasonably straightforward relationship, with people forced to move when their homes are no longer habitable. But in many instances, the effects are counter-intuitive. Water stress, for example, has resulted in decreased north-south migration in Ghana, while similar conditions in Niger were identified as a significant driver of migration. Many studies suggest that environmental crises reduce the number of people on the move as households seek to protect their assets rather than expend capital on the costly process of moving. On the other hand, those that traditionally use migration as a coping strategy can also have their routes cut by extreme events, reducing their ability to withstand the impacts of climate change due to their lack of mobility. These groups may then adopt long-term displacement in place of regular short-term or circulatory patterns of movement (Box 5.3). There is also no evidence to suggest that the distance of migration increases with the severity of impact. Short-term migration has been used as way of diversifying household income, with remittances proving important in communities in degraded rain-fed agriculture regions in particular. Such diversification may become an increasingly important element of adaptation as the incremental impacts of climate change on local livelihoods become evident (Tacoli, 2009). However, reports also suggest that in some circumstances there are more downsides than benefits to this strategy. Research by Tearfund in Ethiopia found that 'ultimately labour migration is negative since remittances rarely reach intended beneficiaries when they are most needed. In addition, the absence of labour power at crucial times of year means that agricultural production is not maximised' (Naess et al., 2010: 15).

The expectation is that – rather than generating large numbers of international refugees – 4°C warming will stimulate population movements over short distances, within state borders, many of which will eventually become permanent (Gemenne, 2011). But given the inherent uncertainty in climate change-induced migration, what can be said of the role for adaptive capacity in helping communities to live with these potential future challenges? What, in fact, does the resilience perspective and adaptive capacity focus proposed by this book mean for this and other possible, profound impacts of climate change, such as recurrent crop failure or dramatically increased water scarcity?

The implication of rapid climate change is that there is an urgent need to work towards processes in which the poor and marginalized gain a voice and are able to take a positive role in planning their future. Migration, for

---

Box 5.3 Changing migration patterns

Traditional nomadic patterns, which were used by pastoralists to cope with droughts, have been modified due to rapidly changing environmental and socio-economic conditions. A similar phenomenon is observed in Bangladesh, where the traditional movement of people from char to char is disrupted by flash floods that are more violent and frequent than they used to be. Thus, it appears that, if the impacts of climate change become more severe, they could disrupt traditional patterns of mobility and people might need to leave their usual place of residence. Migration options would become more limited. In that case, it is expected that the movement would most likely be a long-term or permanent migration instead of a temporary displacement – a trend that has been observed in different countries of South-east Asia and sub-Saharan Africa (most notably Ghana, Vietnam and Bangladesh). In Vietnam, for example, rice farmers usually undertake seasonal labour migration to urban centres during the flooding season, in order to increase and diversify their incomes. Successive floods, however, leading to the destruction of crops, have prompted farmers to migrate permanently in search of a new livelihood.

Source: Gemenne, 2011: 187

example, will not be the most significant or pressing adaptation option for many communities in the short term. But whether migration is an urgent or emerging issue, it constitutes a livelihood transformation, bringing entirely new risks and opportunities. Supporting adaptive capacity in this context draws attention to inclusive decision making, securing rights and building relationships between the host and migrant populations. By providing the opportunity to learn from experiences of mobility as a response to climate change, these processes can better prepare communities for adaptations in response to the changing social, political or environmental context. The alternative is to leave communities to face uncertain futures without support.

Large scale forced displacements, such as in response to sea-level rise, bring the rights issues raised in chapter 4 into sharp focus. Here, supporting adaptive capacity must include working to secure recognition of the rights of migrants. This may require processes to influence policy, or concerted efforts in consensus building fora that ensure the government, the displaced and, where appropriate, the receiving population reach shared decisions as to how best the rights of those affected can be protected. The central issue is empowering communities that may be forced to move to do so in the presence of information about their choices and with the support of the relevant authorities.

Addressing the increasing pressures brought about by climate change will not be straightforward or amenable to prescriptive solutions. It is by no means clear whether, in all circumstances, empowered participation, action and learning will be sufficient to enable the poorest and most vulnerable to achieve the best outcomes in the face of escalating environmental change. However, the perspective offered here is that it will – at the least – be a necessary part of the answer. Significantly, this is a call that is being led by those who are being affected by climate impacts today. For example, many from the low-lying island of Tuvalu have strongly rejected a narrative that robs them of their ability to engage in their own process of change: 'We are not happy to be labelled victims and where is the glory in being titled 'first environmental refugees'? ... Give us real solutions that will empower us to make sustainable choices as we adapt to our changing environment' (Tuvaluan activist, quoted by Farbotko and Lazrus, 2010: 8).

Similar conclusions are being drawn by peasant farmers, who claim the 'right to actively participate in policy design, decision making, implementation, and monitoring of any project, programme or policy affecting their territories' as a prerequisite to sustainable food provision (Article 2(4), Declaration of the Rights of Peasants, 2009). And in Peru, new water laws have prompted growing protest from indigenous and farming communities: 'Transformation of watershed governance will require that those whose vulnerability is greatest speak loudly and maybe carry big sticks. Without their participation, water is unlikely to be governed in ways that reflect its multiple values, and simplification of water rights will reduce watershed capacity to respond to climate change' (Lynch, 2010: 15).

Each of these cases illustrates the pressing need for the poor to have their rights and entitlements recognized as the first step towards power sharing and as a fundamental building block of adaptive capacity. Social movements and alliances of the poor, such as Shack/Slum Dwellers International or the international peasant movement La Via Campesina, and the organizations and activists that make up their membership, are among those leading calls for the poorest to play a defining role in their own futures. In many cases, they have articulated specific demands for action on climate change adaptation (for example, Alliance for Food Sovereignty in Africa, 2009). Development actors that recognize the long-term and profound challenges of adaptation need to ask themselves how they can best support these calls for change.

Adapting development

Chapter 2 opened with a summary of decades of thought and action that have refined our understanding of the adaptive cycle. The insights from this work suggest that there are common elements that need to be present if adaptive capacity is to be supported and resilience built and maintained. This book suggests that it is these elements – power sharing, knowledge and information, and experimentation and testing – that need to be prioritized in development practice. This approach to supporting adaptive capacity opens the door to integrating climate information into community development in a way that builds understanding of climate change at the local level. But it also does much more.

Attention to these dimensions of development work recognizes the inevitability of uncertain futures. They direct action towards creating opportunities for communities to identify important but slowly changing environmental, political and social conditions. In this way, they can help communities to unpack complexity and see emerging threats to well-being. Crucially, by integrating power sharing and experimentation into a learning cycle they also empower communities to make changes to their lives and livelihoods in the presence of enhanced knowledge, and with a greater ability to influence the forces that shape local opportunities for action. In short, they can be supported to build their resilience. But this is an approach that is underpinned by the recognition of the rights and entitlements of the poor. This is a pre-requisite whether communities face transition, transformation or a future of social change and resource competition. Redirecting development focus towards these fundamental issues is also an important purpose of this book.

The strength of approaching climate change through a focus on adaptive capacity lies in recognizing that there are no perfect, off-the-peg solutions to the multiple challenges that communities face. Rather, by focusing attention on processes that include communities in cycles of decision making and learning, the goal of support for adaptive capacity is to set in place the conditions under which the poor and marginalized are increasingly able to adapt to their complex and uncertain futures. Undoubtedly, this perspective challenges development practitioners to think and work differently. It means facing politics and power, empowering poor communities and supporting social transformation. This redefines development practice as working and learning with the poor to enable *them* to take control of *their own* process of development. Supporting adaptive capacity therefore means adapting development. But to do otherwise can only lead to increasing marginalization in a future of unrelenting climate change impacts.

References

Adger, W. N. (2000) 'Social and ecological resilience: are they related?' *Progress in Human Geography* (24)3: 347–364.

Adger, W. N. (2003) 'Social capital, collective action, and adaptation to climate change' *Economic Geography* 79(4): 387–404.

Adger, W. N., Huq, S., Brown, K., Conway, D. and Hulme, M. (2009) 'Adaptation to climate change in the developing world' in E.L.F. Schipper and I. Burton (eds) *The Earthscan Reader on Adaptation to Climate Change*, Earthscan, London, UK.

Adger, W.N., Lorenzoni, I. and O'Brien, K.L. (2009) 'Adaptation now', in W.N. Adger, I. Lorenzoni and K.L. O'Brien (eds) *Adapting to Climate Change, Thresholds, Values, Governance*, pp. 1–22, Cambridge University Press, Cambridge, UK.

Alliance for Food Sovereignty in Africa (2009) 'Challenges African leaders on climate change', Bole Declaration, 25th November 2009, Addis Ababa, Ethiopia.

Anderson, K. and Bows, A. (2011) 'Beyond 'dangerous' climate change: emission scenarios for a new world', *Philosophical Transactions of the Royal Society A* 369: 20–44.

Balstad, R., Russell, R., Gill, V. and Marx, S. (2009) 'Adapting to an uncertain climate on the Great Plains: testing hypotheses on historical populations', in W.N. Adger, I. Lorenzoni and K.L. O'Brien (eds) *Adapting to Climate Change, Thresholds, Values, Governance*, pp. 283–295, Cambridge University Press, Cambridge, UK.

Berkes, F. and Folke, C. (2002) 'Back to the future: ecosystem dynamics and local knowledge', in L.H. Gunderson and C.S. Holling (eds) *Panarchy: Understanding transformations in human and natural systems*, pp. 121–146, Island Press, Washington DC, USA.

Betts, R.A., Collins, M., Hemming, D.L., Jones, C.D., Lowe J.A., and Sanderson, M.G. (2011) 'When could global warming reach 4°C?' *Philosophical Transactions of the Royal Society A* 369: 67–84.

Borrini-Feyerabend, G., Pimbert, M., Farvar, T., Kothari, A. and Renard, Y. (2007) *Sharing power learning-by-doing in co-management of natural resources throughout the world*, Earthscan, London, UK.

Boyd, E., Grist, N., Juhola, S. and Nelson, V. (2009) 'Exploring development futures in a changing climate: frontiers for development policy and practice. Overview to the Special Issue', Development Policy Review 27(6): 659–674.

Boyd, E. and Juhola, S. (2009) 'Stepping up to the climate change: opportunities in re-conceptualising development futures', Journal of International Development 21: 792–804.

Brooks, N. (2003) 'Vulnerability, risk and adaptation: a conceptual framework', *Tyndall Centre for Climate Change Research Working Paper 38*.

Brown K. (2003) 'Integrating conservation and development: a case of institutional misfit', *Frontiers in Ecology and the Environment* 1(9): 479–87.

Buhaug, H., Gleditsch N.P. and Theisen, O.M. (2008) *Implications of Climate Change for Armed Conflict. Social Dimensions of Climate Change*, The World Bank, Washington DC, USA.

Bunce, M., Brown, K. and Rosendo, S. (2010) 'Policy misfits, climate change and cross-scale vulnerability in coastal Africa: how development projects undermine resilience', *Environmental Science and Policy* 13: 485–497.

CAPRi (2010) 'The role of collective action and property rights in climate change', *CAPRi Policy Brief No. 7*.

CARE (2010) 'Community-based adaptation in action', [online] http://www.care.org/getinvolved/advocacy/climatechange/ourwork_adaptation_initiatives.asp [accessed 15 September 2010].

Cash, D.W., Borck, J.C. and Patt, A.G. (2006) 'Countering the loading-dock approach to linking science and decision making', *Science, Technology and Human Values* 31(3): 1–30.

CCAPS (2011) 'The brewing storm? Climate change and African Political Stability Program', *Policy Brief No. 2*, Robert S. Strauss Center for International Security and Law.

Chapin, F.S., Lovecraft, A.L., Zavaleta, E.S., Nelson, J., Robards, M.D., Kofinas, G.P., Trainor, S.F., Peterson, G.D., Huntington, H.P. and Naylor, R.L. (2006) 'Policy strategies to address sustainability of Alaskan boreal forests in response to a directionally changing climate', *Proceedings of the National Academy of Sciences of the United States of America* 103(45): 16637–16643.

Collins K. and Ison R. (2009) 'Jumping off Arnstein's ladder: social learning as a new policy paradigm for climate change adaptation', *Environmental Policy and Governance* 19: 358–373.

Colvin, J., Ballim, F., Chimbuya, S., Everard, M., Goss, J., Klarenberg, G., Ndlovu, S., Ncala, D. and Weston D (2008) 'Building capacity for co-operative governance as a basis for integrated water resource managing in the Inkomati and Mvoti catchments, South Africa', *Water South Africa* 34(6): 681–690.

Commission on Climate Change and Development (2009) *Closing the gaps, Report of the Commission on Climate Change and Development*, Stockholm, Sweden.

Declaration of the Rights of Peasants – Women and Men (2009) 'La Via Campesina, Jakarta', [online] http://viacampesina.net/downloads/PDF/EN-3.pdf [accessed June 2011].

Desai, S., Hulme, M., Lempert, R. and Pielke, R. (2009) 'Climate prediction: a limit to adaptation?', in W.N. Adger, I. Lorenzoni and K.L. O'Brien (eds) *Adapting to Climate Change, Thresholds, Values, Governance*, pp. 64–78, Cambridge University Press, Cambridge, UK.

Eade, D. (2010) 'Capacity building: who builds whose capacity?' in A. Cornwall and D. Eade (eds), *Deconstructing development discourse: buzzwords and fuzzwords*, pp. 203–214, Practical Action Publishing, Rugby, UK.

Ensor, J. and Berger, R. (2009a) *Understanding Climate Change Adaptation*, Practical Action Publishing, Rugby, UK.

Ensor J. and Berger, R. (2009b) 'Community-based adaptation and culture in theory and practice' in W.N. Adger, I. Lorenzoni and K.L. O'Brien (eds) *Adapting to Climate Change, Thresholds, Values, Governance*, pp. 227–239, Cambridge University Press, Cambridge, UK.

Ensor, J. and Weragoda, R. (2009) 'Realizing the adaptive capacity of farming communities in coastal Sri Lanka', *Waterlines*, 28 (3).

Farbotko, C. and Lazrus, H. (2010) '"We are not happy to be labelled victims": contestations and effects of climate refugee narratives', 2nd International Conference: Climate, Sustainability and Development in Semi-Arid Regions, Fortaleza, Brazil.

Few, R., Brown, K. and Tompkins, E. (2007) 'Public participation and climate change adaptation: avoiding the illusion of inclusion', *Climate Policy* 7: 46–59.

Folke, C. (2003) 'Freshwater for resilience: a shift in thinking', *Philosophical Transactions of the Royal Society of London B*: 2027–2036.

Folke, C., Colding, J. and Berkes, F. (2003) 'Synthesis: building resilience and adaptive capacity in social-ecological systems', in F. Berkes, J. Colding and C. Folke (eds) *Navigating Social-Ecological Systems: Building Resilience for Complexity and Change*, pp. 352–387, Cambridge University Press, Cambridge, UK.

Folke, C., Hahn, T., Olsson, P. and Norberg, J. (2005) 'Adaptive governance of social-ecological systems', *Annual Review of Environmental Resources* 30: 441–73.

Gemenne, F. (2011) 'Climate-induced population displacements in a 4°C+ world', *Philosophical Transactions of the Royal Society A* 369: 182–195.

Gine, X., Townsend, R. and Vickery, J. (2007) *Patterns of rainfall insurance participation in rural India*, New York Fed Staff Reports 302.

Goetz, A.M. and Gaventa, J. (2001) 'From consultation to influence: bringing citizen voice and client focus into service delivery', *IDS Working Paper 138*, Institute of Development Studies, Brighton, UK.

Goldstein, B. E. (2009) 'Resilience to surprises through communicative planning' *Ecology and Society* 14(2): 33.

Gore, C. (1997) 'Irreducible social goods and the informational basis of Amartya Sen's capability approach', *Journal of International Development* 9(2): 235–50.

Gready, P. (2008) 'Rights-based approaches to development: what is the value-added?' *Development in Practice*, 18(6): 735–747.

Gready, P. and Ensor, J. (2005a) 'Introduction', in P. Gready and J. Ensor (eds) *Reinventing development? Translating rights-based approaches from theory into practice*, pp. 1–46, Zed Books, London, UK.

Gready, P. and Ensor, J. (eds) (2005b) *Reinventing development? Translating rights-based approaches from theory into practice*, Zed Books, London, UK.

Handmer, J. and Dovers, S. (2009) 'A typology of resilience: rethinking institutions for sustainable development', in E.L.F. Shipper and I. Burton (eds) *The Earthscan reader on adaptation to climate change*, pp. 187–210, Earthscan, London, UK.

Harrison, M., Troccoli, A., Anderson, D.L.T., Mason, S.J., Coughlan, M. and Williams, J.B. (2007) 'A way forward for seasonal climate services', in A. Troccoli, M. Harrison, D.L.T. Anderson and S.J. Mason, *Seasonal Climate: Forecasting and Managing Risk*, pp. 413–425, Springer Academic Publishers, London.

Hartmann, B. (2010) 'Rethinking climate refugees and climate conflict: rhetoric, reality and the politics of policy discourse', *Journal of International Development* 22: 233–246.

Holling, C.S. and Gunderson, L.H. (2002) 'Resilience and adaptive cycles', in L.H. Gunderson and C.S. Holling (eds), *Panarchy: Understanding transformations in human and natural systems*, pp. 25–62, Island Press, Washington DC, USA.

Holling, C.S., Gunderson L.H. and Peterson, G.D. (2002) 'Sustainability and panarchies', in L.H. Gunderson and C.S. Holling (eds), *Panarchy: Understanding transformations in human and natural systems*, Island Press, Washington DC, USA pp. 63–102.

Huq, S. (2007) 'Community based adaption', *IIED Briefing*, International Institute for Environment and Development, London.

ICJ (1981) 'Summary of discussions and conclusions', in International Commission of Jurists (ed.) *Development, Human Rights and the Rule of Law*, Pergamon Press, Oxford, UK.

Jennings, T.L. (2009) 'Exploring the invisibility of local knowledge in decision making: the Boscastle Harbour flood disaster', in W.N. Adger, I. Lorenzoni and K.L. O'Brien (eds), *Adapting to Climate Change, Thresholds, Values, Governance*, pp. 240–254, Cambridge University Press, Cambridge, UK.

Jones, L., Ludi, E. and Levine, S. (2010) 'Towards a characterisation of adaptive capacity: a framework for analysing adaptive capacity at the local level', ODI Background Note, December 2010, ODI, London, UK.

Jonsson, U. (2005) 'A human rights-based approach to programming', in P. Gready and J. Ensor (eds) *Reinventing development? Translating rights-based approaches from theory into practice*, pp. 47–62, Zed Books, London, UK.

Leach, M. (ed.) (2008) 'Re-framing resilience: a symposium report', *STEPS Working Paper* 13, STEPS Centre, Brighton, UK.

Lewins, R., Coupe, S. and Murray, F. (2007) *Voices from the margins: Consensus building and planning with the poor in Bangladesh*, Practical Action Publishing, Rugby, UK.

Lynch, B.D. (2010) 'Equity, vulnerability and water governance: responding to climate change in the Peruvian Andes', 2nd International Conference: Climate, Sustainability and Development in Semi-Arid Regions, Fortaleza, Brazil.

Mander, H. (2005) 'Rights as struggle – towards a more just and humane world', in P. Gready and J. Ensor (eds), *Reinventing development? Translating rights-based approaches from theory into practice*, pp. 233–253, Zed Books, London, UK.

Marshall, N.A., Marshall, P.A., Tamelander, J., Obura, D., Malleret-King, D. and Cinner, J.E. (2009) 'A framework for social adaptation to climate change; sustaining tropical coastal communities and industries', IUCN Gland, Switzerland.

Min, S-K., Zhang, X., Zwiers, F.W. and Hegerl, G.C. (2011) 'Human contribution to more-intense precipitation extremes', *Nature* 470: 378–381.

Mitchell, T. and Tanner, T.M. (2006) 'Adapting to climate change: challenges and opportunities for the development community. IDS and Tearfund, Teddington, UK.

Molyneux, M. and Lazar, S. (2003) *Doing the rights thing: rights-based development and Latin American NGOs*, ITDG Publishing, Rugby, UK.

Moser, S.C. (2009) 'Whether our levers are long enough and the fulcrum strong? Exploring the soft underbelly of adaptation decisions and actions', in W.N. Adger, I. Lorenzoni and K.L. O'Brien (eds) *Adapting to Climate Change, Thresholds, Values, Governance*, pp. 313–334, Cambridge University Press, Cambridge, UK.

Mulvany, P., O'Riordan, B. and Wedgwood, H. (1995) 'Taking root ... gaining ground: diversity in food production for universal food security', Paper presented to the Development Studies Association, ITDG, Rugby, UK.

Murwira, K., Wedgewood, H. Watson, C., and Win, W.J. with Tawney, C. (2000) *Beating hunger. The Chivi experience,* Intermediate Technology Publications, Rugby, UK.

Naess, L.O., Sullivan, M., Khinmaung, J., Crahay, P. and Otzelberger, A. (2010) 'Changing climates changing lives', Tearfund, UK.

Nelson, D.R., Adger, W.N. and Brown, K. (2007) 'Adaptation to environmental change: contributions of a resilience framework', *Annual Review of Environment and Resources*, 32: 395–419.

Nelson, P.J. and Dorsey E. (2003) 'At the Nexus of Human Rights and Development: New Methods and Strategies of Global NGOs' *World Development* 31(12): 2013–2026.

New, M., Liverman, D., Schroder, D. and Anderson, K. (2011) 'Four degrees and beyond: the potential for a global temperature increase of four degrees and its implications', *Philosophical Transactions of the Royal Society* A 369:6–19.

O'Brien, K., Hayward, B. and Berkes, F. (2009) 'Rethinking social contracts: building resilience in a changing climate', *Ecology and Society* 14(2): 12.

OHCHR (1990) The nature of States parties obligations (Art. 2, par.1): General Comment 3 on the Convention on Economic and Social and Cultural Rights 12/14/1990.

OHCHR (2009) Report of the Office of the United Nations High Commissioner for Human Rights on the Relationship Between Climate Change and Human Rights A/HRC/10/61 Human Rights Council Tenth Session, Geneva: United Nations.

Oppenheimer, M., O'Neill B.C., Webster M. and Agrawala S. (2007) 'The limits of consensus', *Science*, 317: 1505–6.

Osbahr, H., Twyman, C., Adger, W.N. and Thomas, D.S.G. (2008) 'Effective livelihood adaptation to climate change disturbance: scale dimensions of practice in Mozambique', *Geoforum* 39: 1951–1964.

Patt, A. (2005) 'Effects of seasonal climate forecasts and participatory workshops among subsistence farmers in Zimbabwe', *Proceedings of the National Academy of Sciences of the United States of America* 102(35): 12623–12628.

Patt, A. (2008a) 'How does using seasonal forecasts build adaptive capacity?', in *Living With Climate Change: Are There Limits to Adaptation?*, Conference Proceedings 7–8 February 2008, pp. 62–67, Royal Geographical Society, London.

Patt, A. (2008b) *Personal communication: email 6 June 2008*, on record with the author.

Pelling, M. (2010) *Adaptation to Climate Change: From Resilience to Transformation* Routledge, London, UK.

Peterson, G. (2000) 'Political ecology and ecological resilience: an integration of human and ecological dynamics', *Ecological Economics* 35: 323–336.

Pettengell, C. (2010) 'Climate change adaptation: enabling people living in poverty to adapt', Oxfam Research Report, Oxfam, UK.

Phillips, A. (2003) 'Turning ideas in their heads', in H. Jaireth and D. Smyth (eds), *Innovative governance*, pp. 1–28, Ane Books, Delhi.

Pimbert, M.P. and Wakeford, T. (2002) 'Prajateerpu. A citizens' jury/scenario workshop for food and farming in Andhra Pradesh, India', International Institute for Environment and Development, Institute of Development Studies, Andhra Pradesh Coalition in Defence for Diversity, University of Hyderabad and All India National Biodiversity Strategy and Action Plan, IIED, London, UK.

Pimbert, M.P. and Wakeford, T. (2003) 'Prajateerpu, power and knowledge: the politics of participatory action research in development. Part 1: Context, process and safeguards', *Action Research*, 1(2): 184–207.

Pimbert, M. (2007) *Transforming knowledge and ways of knowing for food sovereignty*, IIED Publications, London, UK.

Plumber, R. and Armitage, D. (2010) 'Integrating perspectives on adaptive capacity and environmental governance', in D. Armitage and R. Plumber (eds), *Adaptive Capacity and Environmental Governance*, pp. 1–22, Springer Series on Environmental Governance, Springer, Heidelberg, Germany.

Practical Action (2011) 'Training for transformation participant reader', Practical Action Southern Africa internal document, Harare, Zimbabwe.

Ribot, J. (2009) 'Vulnerability does not fall from the sky: toward multiscale, pro-poor climate policy', in R. Mearns and A. Norton (eds), *Social Dimensions of Climate Change: Equity and Vulnerability in a Warming World*, pp. 47–74,The World Bank, Washington DC, USA.

Right to Food Campaign (2011) http://www.righttofoodindia.org/links/articles_intro. html [accessed 1 April 2011].

Röhr, U. (2007) 'Gender, climate change and adaptation. Introduction to the gender dimensions', Background Paper prepared for the Both Ends Briefing Paper *Adapting to climate change: How local experiences can shape the debate*, August 2007.

Saferworld (2010) 'Climate change and conflict: a framework for analysis and action'. Working paper, Saferworld, London, UK.

Schumpeter, J.A. (1950) *Capitalism, socialism and democracy*, Harper and Row, New York, USA.

Scoones, I. and Thompson, J. (eds) (2009) *Farmer First Revisited*, Practical Action Publishing, Rugby, UK.

Sen, A. (1999) *Development as Freedom*, Oxford University Press, Oxford, UK.

Smit, B. and Wandel, J. (2006) 'Adaptation, adaptive capacity and vulnerability', *Global Environmental Change* 16: 282–292.

Smith, L. (2009) 'One two three more: challenges to describing a warmer world', 4 Degrees and Beyond International Climate Conference, Oxford, UK, 28–30 September 2009.

Smith, D. and Vivekananda, J. (2009) 'Climate change, conflict and fragility: understanding the linkages, shaping effective responses', International Alert, London, UK.

Stafford Smith, M., Horrocks, L., Harvey, A. and Hamilton C. (2011) 'Rethinking adaptation for a 4°C world', *Philosophical Transactions of the Royal Society A* 369: 196–216.

Stainforth, D.A., Allen, M.R., Tredger, E.R. (2007) 'Confidence, uncertainty and decision-support relevance in climate predictions', *Philosophical Transactions of the Royal Society A*, 365: 2145–2161.

Tacoli, C, (2009) 'Crisis or adaptation? Migration and climate change in a context of high mobility', in J.M. Guzman, G. Martine, G. McGanahan, D. Schensul and C. Tacoli (eds), *Population dynamics and climate change*, pp. 104–118, UNFPA and IIED, London, UK.

Taha, A., Lewins, R., Coupe, S. and Peacocke, B. (2010) 'Consensus building with Participatory Action Plan Development', Practical Action Facilitators Guide, June 2010.

Terry, G. (2009) 'Introduction', in G. Terry (ed.) *Climate change and gender justice*, Practical Action Publishing, Rugby, UK.

Tomas, A. (2005) 'Reforms that benefit poor people – practical solutions and dilemmas of rights-based approaches to legal and justice reform', in P. Gready and J. Ensor (eds), *Reinventing development? Translating rights-based approaches from theory into practice*, pp. 171–184, Zed Books, London, UK.

Tompkins, E.L. and Adger, W.N. (2004) 'Does adaptive management of natural resources enhance resilience to climate change?', *Ecology and Society*, 9.

Trosper, R. L. (2003) 'Resilience in pre-contact Pacific Northwest social ecological systems', *Conservation Ecology* 7(3): 6.

Tschakert, P. (2007) 'Views from the vulnerable: understanding climatic and other stressors in the Sahel', *Global Environmental Change* 17: 381–396.

UNDP (2010a) 'Gender, climate change and community-based adaptation a guidebook for designing and implementing gender-sensitive community-based adaptation programmes and projects', UNDP, New York, USA.

UNDP (2010b) 'Community-based adaptation to climate change', UNDP, New York, USA.

Uvin, P. (2004) *Human Rights and Development*, Kumarian Press, Bloomfield, CT, USA.

Uvin, P. (2010) 'From the right to development to the rights-based approach: how 'human rights' entered development', in A. Cornwall and D. Eade (eds) *Deconstructing development discourse: buzzwords and fuzzwords*, pp. 163–174, Practical Action Publishing, Rugby, UK.

Vincent, K. (2007) 'Uncertainty in adaptive capacity and the importance of scale', *Global Environmental Change* 17: 12–24.

Walker, B, Holling, C.S, Carpenter, S.R. and Kinzig, A. (2004) 'Resilience, adaptability and transformability in social-ecological systems', *Ecology and Society* 9(2): 5.

Walker, B. and Salt, D. (2006) *Resilience thinking: sustaining ecosystems and people in a changing world*, Island Press, Washington DC, USA.

Wasserman, S. and Faust, K. (1994) *Social Network Analysis: Methods and Applications*. Cambridge University Press, Cambridge.

Yates, J. S. (2011, forthcoming) 'Institutional complexity in governing the scalar politics of livelihood adaptation in rural Nepal', Paper presented at the Annual Meeting of the Association of American Geographers, Washington State Convention Center, Seattle, Washington.

Index

absorbing capacity 6–8, 15–6, 24–5
access to
    information 32, 40, 43–4, 46, 49, 57
    resources 10, 26, 32, 90, 92
accountability 52–6, 81, 83, 86
actions *see* adaptation; collective; development
activism *see* political
adaptation
    actions 9–10, 23, 32–4, 45, 55, 87
    decision making 37–8, 49
    measures 5, 8, 33, 42, 91
    planning 3–6
    see also climate change; community-based
adapting development practice 51–83
adaptive capacity 8, 13–5, 23, 31, 51, 56
    see also capacity; support for adaptive capacity
adaptive cycles 18, 20–5
    see fore loop; release phase
Africa 88–90, 93–5
agricultural 5–6, 8–9, 42, 73–5, 80, 93
Ajaweed, Sudan 91–2
alternative
    strategies 6, 11, 25
    technologies 46–7, 74, 79
Andhra Pradesh, India 46, 72–3
assets 2, 6, 22–5, 31–3, 47, 92
awareness raising 1, 7, 53–6, 59–60, 71, 88

Bangladesh 58–9, 63, 68–71, 75, 79–82, 91–3
    see charlands
barriers 8, 23, 40, 53, 57, 64
'business as usual' approach 2, 9, 37, 88

capacity
    building 8–9, 56, 67–8, 83
    see also absorbing capacity; technical capacity
capital 21–3, 26–7, 31–2, 65, 92
    see also social
charlands, Bangladesh 58, 63–7

Chivi, Zimbabwe 76–8, 83
citizens' jury 73, 80
climate
    information 44–5, 57, 88, 95
    modelling 3–5, 45
    science 2–3, 5, 19
climate change
    adaptation 28, 42, 45, 95
    debates 89–91
    impacts 3, 5, 9–13, 17, 24–8, 85–96
    predictions 2–4, 44–5
    see also Swedish Commission on Climate Change
coastal 6, 28, 69
collective action 36, 59, 66–7
co-management 27, 40–1, 58, 62–3, 68–71, 83
command and control 19, 27
committees 61, 65–7, 71–2, 82, 90–1
community-based 27, 59, 62, 71, 74
    adaptation 1–17
competition
    see resource
complex systems 13, 18–20, 29, 33–5, 41, 52
complexity 5, 11–4, 18–26, 49, 85–7, 95
    and uncertainty 31–3, 40–45
conflict 12, 41, 55, 89–92
connectedness 21–3, 59
consensus building 4, 51, 58–9, 69–70, 94
    approaches 12, 58, 80, 91
conservation 20–5, 31, 78
crop
    failures 88, 93
    varieties 6, 8, 23, 26, 78, 87
cross-scale 23, 26, 28
culture 2, 10, 39, 52
    see also ESCRs

decision making *see* adaptation; local
development
    actions 10–1, 32, 40, 43, 48, 53
    practitioners 11–2, 15, 75–6, 81, 96
    see also adapting development practice; human; PAPD; PTD

DIPs (deliberative and inclusive processes) 72–4, 80
disaster 8, 87, 90
    risk reduction (DRR) 71
discrimination 40–1, 53, 58
displacement 73, 88, 93
disputes 65, 89–92
disturbance 13–4, 16, 20, 25
    see also external
diversity 6–8, 19, 22–4, 31–2, 47, 80
drought 6, 20, 23–4, 26, 42, 90
    frequency 17, 45–6, 76
    tolerance 8, 79
duty bearers 10, 52, 54–6, 61, 83

ecological resilience 14–5, 18, 89
    see also social-ecological
economic 10, 13, 23, 38–42, 52–4, 68–9
    see also socio-economic; ESCRs
ecosystems 10, 14, 18–20, 28
education 32, 37, 54
El Niño 4–5
emissions 3–5, 15, 86, 88
empowering communities 2, 12, 17, 49, 53–5, 94–6
entitlements 32, 40–1, 62–7, 70, 86, 95
    see also legitimate
environmental change 2, 10, 26, 47, 89, 92–4
equilibrium 14, 19, 85, 88
equity 38–41, 51, 55, 77, 81, 91
erosion 6, 58
ESCRs (economic, social and cultural rights) 56–8
Ethiopia 90, 93
exclusion 10, 52, 54
experimentation 22–5, 41
    and testing 11–2, 34–5, 46–51, 74–81, 88, 95
exposure visits 74, 76–9
extension 48, 59, 61, 78
external
    actors 74, 77, 82
    disturbances 13–4, 16, 85
    influences 18, 21, 33
extreme weather 2–3, 10, 45, 86, 89, 93

facilitation 41–2, 47–9, 58–9, 68–9, 78, 83
farming 13, 37, 44, 72–5, 94

feedback 61, 77, 79
financial 31–2, 38, 41, 61, 65–6, 82
fisheries 13, 37, 59, 61–3, 66, 82
fishing 14, 24, 61–3, 65
flooding 1, 5–6, 93
food 15, 22, 32, 45, 56, 94–5
    security 5–6, 26, 63, 73, 76, 87–9
fore loop 21–3, 25
forecasting 4, 9, 26, 45
forestry 37, 69
function 11, 14–7, 19–20, 23–4, 29, 33–4
funding 2, 61, 80
future uncertainty 3, 5, 12, 31, 85–9, 92–3

gender 2, 10, 37, 52–4, 77
Ghana 92–3
global warming 2, 12, 86, 88
governance 23–6, 32, 63, 68, 90, 94
    structures 10, 35, 38, 49, 81–2
greenhouse gas 3–5, 86
gusthi (kinship groups) 65–7

hazards 6–7, 71
health 5, 20, 32, 57, 87
human
    development 9, 18, 32
    rights 52–3, 56–7, 81

impacts see climate change
income 7, 61, 63, 93
incremental changes 12–3, 16–7, 24, 28, 45, 87–8
inclusivity 27–8, 42, 49, 91, 94
    see also DIPs
India 56, 59, 73
    see also Andhra Pradesh
indigenous 38, 91, 94
    knowledge 9, 19, 24, 44, 73, 80
information
    gathering 60–4, 67, 79–80
    see also access to; climate; knowledge and information
insurance 8, 19, 32, 46
interdependencies 69–70
interest groups 26, 58, 62–5, 67–8, 73, 82
irrigation 23, 69

justice 52, 58, 90–1

Kenya 26, 90–2
knowledge and information 11–2, 34–5, 43–51, 67, 82–5, 88
    see also indigenous; local; skills

labour 65, 73, 93
lakes 5, 14, 19–20
land 6, 41–2, 55–8, 63–7, 69–75, 78
law 12, 56–8, 72, 86
learning *see* shared learning; social
legal 38, 41, 55–8, 64
legitimate 36, 40, 49, 71
    entitlements 52–4, 63–4
lifetime 18, 76, 87–8
livelihood
    changes 2, 12, 15, 29, 34, 67
    strategies 3, 6–7, 12, 44–6,
    transformations 15, 26, 31, 86, 88–9, 94–5
    transitions 86–8
    see also rural
livestock 17, 42, 46, 73
local
    decision making 2, 39, 88, 92
    knowledge 1, 25, 44–5, 67, 74–6, 79–82
long term trends 4–5, 14, 33, 46, 81, 93–5

marginalization 10, 52, 54, 90, 96
meteorological 6, 26, 80
migration 12, 89–94
Mozambique 26, 28, 42
Mvoti, South Africa 68–70

natural resource 18, 31, 36–42, 57–9, 63, 89
negotiation 40–1, 44, 61–4, 68, 70, 81
Nepal 39, 46, 71
networks 25–7, 43–4, 54, 61–2, 85, 91–2
    of relationships 2, 21, 25, 48
    see also gusthi; social
NGOs 26–8, 60, 66, 82–3
Niger 17, 55, 92
'no regrets' approach 6–7, 87
nutrient run-off 14, 20

ocean 5, 24, 86

panarchy 23–5
PAPD (Participatory Action Plan Development) 51, 58–69, 74–6, 79–82, 91
    see STEPS analysis
participation 12, 25–8, 42–4, 52–9, 79–82, 94
participatory
    research 74, 78–9
    see PAPD; PTD
pastoralists 17, 23, 90, 93
pathways 16–7, 25, 59, 69
pattern analysis 59, 61–2
planting 6–7, 78, 87
political
    activism 53–6
    context 25, 35, 41–2, 53, 67–8, 71
    power 38, 67
    processes 10, 54–5
power sharing 11–2, 41–9, 51–3, 67–72, 95
    approaches 42–3, 70, 76, 79–80
    arrangements 34, 53, 58–62, 70, 82–3, 85
    relationships 12, 34–5, 48–9, 51–2, 79, 82
precipitation 3, 5, 86
predictions *see* climate change
processes of change 2, 9, 13, 15, 31–3
projections 4–5, 31, 86
PTD (participatory technology development) 12, 47–8, 74–82

rainfall 2–7, 10, 26, 45, 86–7, 90
rapid growth 20–2, 24–5
release phase 20–5
reorganization 20–5
resilience 13, 85–9, 93–5
    thinking 11, 18–9, 23–4, 28–34, 39, 54
    see also social
resolution *see* conflict; *salish*
resource
    competition 12, 87, 89, 90, 95
    management 23, 26–7, 46, 59–60, 63, 67–70
    see access to; natural resource
responses 1, 11, 16–8, 27, 43–9, 90

responsibility 10, 52–3, 55–6, 62, 71, 83
rights 10, 24–6, 40–2, 48, 90–5
    see also human
rights based 2, 51–8, 72, 83
risk 1–2, 19, 37, 41, 45–8, 83
    see also disaster
rules 23, 39, 41, 63–4
rural 7, 46, 58, 69–75, 81, 92
    livelihoods 10, 14

saline-affected 5, 7–9, 75
salish 65–7, 91
sea levels 3–5, 45, 86, 88–9, 92, 94
seasonal 4–5, 10, 13, 45, 92–3
seed fairs 6, 48, 66, 75, 78
shared learning 35, 42–9, 67–8, 74, 78, 82–3
shocks 6, 8, 15–6, 37
short-term patterns 4, 6–7, 9, 20, 45, 87
skills 8–9, 21–2, 27, 41–2, 65, 78
    and knowledge 47, 73–6
slow variables 19–20, 23–4
social
    capital 27, 45, 63
    learning 27–8, 34–5, 68
    networks 31–2, 35–7, 40, 46
    resilience 14–5
    see also ESCRs
social-ecological systems 14–6, 19–23, 28–9, 31–3, 48–9, 85
socio-economic 25–6, 32, 56–7, 81–3, 92–3
South Africa see Mvoti
Sri Lanka 75
stability 18–9, 22–3, 25
stakeholders 11, 26–7, 34–5, 41–4, 47–51, 91
    multiple 62, 70, 85–8
    primary 59–60, 62, 64, 70–5
    secondary 59–62, 64, 68, 79–82
STEPS analysis 41, 61–3, 65, 67
Sudan 26, 59, 90–2
    see also Ajaweed
support for adaptive capacity 11, 32–4, 51, 62, 81–6, 91
sustainability 24, 61
sustainable 10, 28, 37, 88, 94
Swedish Commission on Climate Change 9, 25, 32, 44

technical capacity 47–8, 74, 76–7
technologies 1, 73–4, 76–82
    see also alternative
technology 6, 51
    choice 47–8, 74, 76, 79
    see also PTD
temperature 3–5, 10, 20, 45
testing 25, 29, 85
    see also experimentation
thresholds 18–20, 23, 34, 48
trade 17, 25, 43, 73, 78
traditional 6, 12, 24–7, 44, 75–6, 91–3
training 6, 19, 48, 61, 74–9, 83
transformation 10, 16–7, 22, 42, 69, 83
    see also livelihood
transitions see livelihood
transparency 64–5, 81, 91
turbidity 14, 19–20

uncertainty 2–9, 19, 22–4, 29, 49, 81
    see complexity; future uncertainty
U-process 69–70, 80

Vietnam 59, 93
violence 65–6, 89–91
Vision 2020 72–3
voice 27, 34, 55, 62, 72–3, 82–5, 93
vulnerability 3, 22, 36–7, 53, 91, 94
    reduction 5–9, 16–7, 44–5

water 1, 6–8, 14, 20, 26, 32
    body 63–8
    management 37, 59, 68–9, 77–8, 94
    scarcity 87–93
wealth 32, 47
weather 4–8, 15, 20, 43
    see also extreme weather
well-being 11–3, 23–4, 31, 73, 89–90, 95
women 36–7, 42, 60–2, 64–7, 73, 75
workshops 44, 69, 71, 73
worldviews 11, 42, 49

yields 7–8, 23, 58, 88

Zimbabwe 17, 44, 71, 75–6
    see Chivi